Student Support Book
with answers

Foundation MATHEMATICS for OCR GCSE

Tony Banks and David Alcorn

Causeway Press

Pearson Education Limited
Edinburgh Gate
Harlow
Essex
CM20 2JE
England

ISBN-13: 978-1-4058-3503-9
ISBN-10: 1-4058-3503-6

Exam questions
Past exam questions, provided by the *Oxford Cambridge and RSA Examinations*, are denoted by the letters OCR. The answers to all questions are entirely the responsibility of the authors/publisher and have neither been provided nor approved by OCR.

Page design
Billy Johnson

Reader
Barbara Alcorn

Artwork
David Alcorn

Cover design
Raven Design

Typesetting by Billy Johnson, San Francisco, California, USA

Printed and bound by Scotprint, Haddington, Scotland

preface

This book provides detailed revision notes, worked examples and examination questions to support students in their preparation for the new two-tier GCSE Mathematics examinations for the OCR Specifications – Foundation Tier.

The book has been designed so that it can be used in conjunction with the companion book ***Foundation Mathematics for OCR GCSE*** or as a stand-alone revision book for self study and provides full coverage of the new OCR Specifications for the Foundation Tier of entry.

In preparing the text, full account has been made of the requirements for students to be able to use and apply mathematics in written examination papers and be able to solve problems in mathematics both with and without a calculator.

The detailed revision notes, worked examples and examination questions have been organised into 40 self-contained sections which meet the requirements of the National Curriculum and provide efficient coverage of the specifications.

Sections 1 - 11	Number
Sections 12 - 20	Algebra
Sections 21 - 33	Shape, Space and Measures
Sections 34 - 40	Handling Data

At the end of the sections on Number, Algebra, Shape, Space and Measures and Handling Data, section reviews are provided to give further opportunities to consolidate skills.

At the end of the book there is a final examination questions section with a further compilation of exam and exam-style questions, organised for non-calculator and calculator practice, in preparation for the exams.

Also available ***Without Answers: (ISBN: 1-405835-02-8)***
The book has been designed so that it can be used in conjunction with the companion book
Foundation Mathematics for OCR GCSE (ISBN: 1-405831-41-3)

contents

Number

Sections 1 - 11

1 Whole Numbers 1
2 Decimals 3
3 Approximation and Estimation 5
4 Negative Numbers 8
5 Fractions 10
6 Working with Number 12
7 Percentages 15
8 Time and Money 17
9 Personal Finance 19
10 Ratio and Proportion 21
11 Speed and Other Compound Measures 23

Section Review - Number

Non-calculator Paper 25
Calculator Paper 28

Algebra

Sections 12 - 20

12 Introduction to Algebra 31
13 Solving Equations 33
14 Further Equations 34
15 Formulae 36
16 Sequences 38
17 Coordinates and Graphs 40
18 Using Graphs 42
19 Inequalities 45
20 Quadratic Graphs 46

Section Review - Algebra

Non-calculator Paper 47
Calculator Paper 50

Shape, Space and Measures

Sections 21 - 33

21 Angles 53
22 Triangles 55
23 Symmetry and Congruence 57
24 Quadrilaterals 59
25 Polygons 61
26 Direction and Distance 63
27 Circles 65
28 Areas and Volumes 67
29 Loci and Constructions 71
30 Transformations 73
31 Enlargements and Similar Figures 76
32 Pythagoras' Theorem 78
33 Understanding and Using Measures 80

Section Review - Shape, Space and Measures

Non-calculator Paper 82
Calculator Paper 85

Handling Data

Sections 34 - 40

34 Collection and Organisation of Data 88
35 Pictograms and Bar Charts 91
36 Averages and Range 93
37 Pie Charts and Stem and Leaf Diagrams 96
38 Time Series and Frequency Diagrams 98
39 Scatter Graphs 101
40 Probability 103

Section Review - Handling Data

Non-calculator Paper 106
Calculator Paper 109

Exam Practice

Non-calculator Paper 112
Calculator Paper 116

Answers 120

Index 136

SECTION 

Whole Numbers

What you need to know

- You should be able to read and write numbers expressed in figures and words.

  Eg 1 The number 8543 is written or read as, "eight thousand five hundred and forty-three".

- Be able to order whole numbers.

  Eg 2 Write the numbers 17, 9, 35, 106 and 49 in ascending order.

  9, 17, 35, 49, 106

Smallest number	ascending order →	Largest number
Largest number	descending order →	Smallest number

- Be able to recognise the place value of each digit in a number.

  Eg 3 In the number 5384 the digit 8 is worth 80, but in the number 4853 the digit 8 is worth 800.

- Use mental methods to carry out addition and subtraction.
- Know the Multiplication Tables up to 10×10.

×	1	2	3	4	5	6	7	8	9	10
1	1	2	3	4	5	6	7	8	9	10
2	2	4	6	8	10	12	14	16	18	20
3	3	6	9	12	15	18	21	24	27	30
4	4	8	12	16	20	24	28	32	36	40
5	5	10	15	20	25	30	35	40	45	50
6	6	12	18	24	30	36	42	48	54	60
7	7	14	21	28	35	42	49	56	63	70
8	8	16	24	32	40	48	56	64	72	80
9	9	18	27	36	45	54	63	72	81	90
10	10	20	30	40	50	60	70	80	90	100

- Be able to: multiply whole numbers by 10, 100, 1000, …
  multiply whole numbers by 20, 30, 40, …
  divide whole numbers by 10, 100, 1000, …
  divide whole numbers by 20, 30, 40, …

  Eg 4 Work out. (a) $75 \times 100 = 7500$ (b) $42 \times 30 = 42 \times 10 \times 3 = 420 \times 3 = 1260$

  Eg 5 Work out. (a) $460 \div 10 = 46$ (b) $750 \div 30 = (750 \div 10) \div 3 = 75 \div 3 = 25$

- Use non-calculator methods for addition, subtraction, multiplication and division.

  Eg 6 476 + 254

$$\begin{array}{r} 476 \\ +\,254 \\ \hline 730 \\ {\scriptstyle 1\,1} \end{array}$$

  Eg 7 374 − 147

$$\begin{array}{r} 3\,\overset{6}{\not7}\,\overset{1}{4} \\ -\,147 \\ \hline 227 \end{array}$$

Addition and Subtraction
Write the numbers in columns according to place value.
You can use addition to check your subtraction.

  Eg 8 324 × 13

$$\begin{array}{r} 324 \\ \times\quad 13 \\ \hline 972 \\ +\,3240 \\ \hline 4212 \\ {\scriptstyle 1\,1} \end{array}$$

  Eg 9 343 ÷ 7

$$\begin{array}{r} 49 \\ 7\overline{)343} \\ 28 \\ \hline 63 \\ 63 \\ \hline 0 \end{array}$$

Long division
→ ÷ (Obtain biggest answer possible.)
Calculate the remainder.
Bring down the next figure and repeat the process until there are no more figures to be brought down.

Long multiplication
Multiply by the units figure, then the tens figure, and so on.
Then add these answers.

- Know the order of operations in a calculation.

First	Brackets and Division line
Second	Divide and Multiply
Third	Addition and Subtraction

Eg 10 $4 + 2 \times 6 = 4 + 12 = 16$

Eg 11 $9 \times (7 - 2) + 3 = 9 \times 5 + 3 = 45 + 3 = 48$

Exercise 1

Do not use a calculator for this exercise.

1 (a) Write in figures one hundred and eleven thousand, nine hundred and ninety-four.
(b) Write 14 076 in words.
OCR

2 (a) In the number 23 547 the 4 represents 4 tens. What does the 3 represent?
(b) Write the numbers 85, 9, 23, 117 and 100 in order, largest first.

3

3	6	50	57	65	71

Copy and complete these sentences using numbers from the box.
You can use a number more than once.
(a) .... are all the odd numbers.
(b) .... can be divided by five exactly.
(c) .... can be divided by ten exactly.
(d) ... and ... add up to 60.
OCR

4 (a) What must be added to 19 to make 100?
(b) What are the missing values?
(i) $100 - 65 = \square$ (ii) $12 \times \square = 1200$ (iii) $150 \div \square = 15$
(c) Work out. (i) $769 + 236$ (ii) $400 - 209$ (iii) $258 - 75$

5 The chart shows the distances in kilometres between some towns.

Bath			
104	Poole		
153	133	Woking	
362	452	367	Selby

Tony drives from Poole to Bath and then from Bath to Selby.
(a) How far does Tony drive?

Jean drives from Poole to Woking and then from Woking to Selby.
(b) Whose journey is longer?
How much further is it?

6 One box holds 18 Christmas cards.
(a) Peter buys 6 boxes.
How many cards are there altogether in 6 boxes?
(b) Mary spends £15 on the boxes.
How many cards does she get?

OCR

7 (a) Complete these calculations.
(i) $26 \times 3 = \ldots$ (ii) $267 \div 3 = \ldots$ (iii) $75 - \ldots = 52$
(b) Look at these number cards.

2	3	5	6	7	8

Use all six number cards to make the largest possible answer to this sum.
Copy and complete the empty cards and give the answer.

OCR

8 (a) By using each of the digits 8, 5, 2 and 3, write down:
(i) the smallest four-digit number, (ii) the largest four-digit odd number.
(b) What is the answer when you subtract the smallest number from the largest odd number?

9 (a) Write down the next line in this number pattern.

$$5 \times 7 = 35$$
$$55 \times 7 = 385$$
$$555 \times 7 = 3885$$
$$5555 \times 7 = 38885$$
$$\times 7 =$$

(b) Carol says this is a line from the same number pattern.

$$55555555 \times 7 = 3888888885$$

Explain why 3888888885 is wrong. OCR

10 Work out. (a) $2655 \div 9$ (b) 417×28 OCR

11 Last year Mr Alderton had the following household bills.

Gas	£364	Electricity	£158	Telephone	£187
Water	£244	Insurance	£236	Council Tax	£983

He paid the bills by 12 equal monthly payments.
How much was each monthly payment?

12 Calculate the cost of 24 rail tickets at £128 each. OCR

13 Kate bought some new doors for her house. They cost £64 each. She paid £448.
How many doors did she buy? OCR

14 A bus started from the bus station with 30 passengers on board.
At the first stop 5 got off and some got on. At the second stop no one got off, but 6 got on.
There were then 38 passengers on the bus. How many got on at the first stop? OCR

15 A supermarket orders one thousand two hundred tins of beans.
The beans are sold in boxes of twenty-four. How many boxes of beans are ordered?

16 Work out. (a) $6 + 4 \times 3$ (b) $96 \div (3 + 5)$ (c) $2 \times (18 - 12) \div 4$

17 Simon is 8 kg heavier than Matt. Their weights add up to 132 kg. How heavy is Simon?

18 A roll of wire is 500 cm long. From the roll, Debra cuts 3 pieces which each measure 75 cm and 4 pieces which each measure 40 cm. How much wire is left on the roll?

19 There are 232 children in Joshua's school. There are 18 more girls than boys.
How many girls are in Joshua's school? OCR

20 Car Hire Co. have the following cars available to rent.

Model	Corsa	Astra	Zafra
Number of cars	10	12	6
Weekly rental	£210	£255	£289

Work out the total weekly rental when all the cars are hired.

21 Lauren works in a car factory. She inspects 14 cars a day. Last year she inspected 3052 cars.
For how many days did she work last year?

22 Look at these calculations, they show the beginning of a number pattern.

① $1 = \frac{1 \times 2}{2} = 1$ ② $1 + 2 = \frac{2 \times 3}{2} = 3$ ③ $1 + 2 + 3 = \frac{3 \times 4}{2} = 6$

(a) Complete the next calculation in the pattern: ④ $1 + 2 + 3 + 4 = \ldots = \ldots$
(b) Hence, work out the sum of the first 100 whole numbers.

SECTION 2 Decimals

What you need to know

- You should be able to write decimals in order by considering place value.

Eg 1 Write the decimals 4.1, 4.001, 4.15, 4.01, and 4.2 in order, smallest first.
4.001, 4.01, 4.1, 4.15, 4.2

- Be able to use non-calculator methods to add and subtract decimals.

Eg 2 2.8 + 0.56

```
   2.8
 + 0.5 6
 -------
   3.3 6
   1
```

Eg 3 9.5 − 0.74

```
   8 14 1
   9.5 0
 − 0.7 4
 -------
   8.7 6
```

Addition and Subtraction
Keep the decimal points in a vertical column.
9.5 can be written as 9.50.

- You should be able to multiply and divide decimals by powers of 10 (10, 100, 1000, ...)

Eg 4 Work out. (a) $6.7 \times 100 = 670$ (b) $0.35 \times 10 = 3.5$ (c) $5.4 \div 10 = 0.54$ (d) $4.6 \div 100 = 0.046$

- Be able to use non-calculator methods to multiply and divide decimals by other decimals.

Eg 5 0.43×5.1

```
       0.4 3   (2 d.p.)
 ×       5.1   (1 d.p.)
 -----------
         4 3 ← 43 × 1
 +   2 1 5 0 ← 43 × 50
 -----------
     2.1 9 3   (3 d.p.)
```

Multiplication
Ignore the decimal points and multiply the numbers.
Count the total number of decimal places in the question.
The answer has the same total number of decimal places.

Eg 6 $1.64 \div 0.2$

$\frac{1.64}{0.2} = \frac{16.4}{2} = 8.2$

Division
It is easier to divide by a whole number than by a decimal.
So, multiply the numerator and denominator by a power of 10 (10, 100, ...) to make the dividing number a whole number.

- Be able to use decimal notation for money and other measures.

The metric and common imperial units you need to know are given in Section 33.

- Be able to change decimals to fractions.

Eg 7 (a) $0.2 = \frac{2}{10} = \frac{1}{5}$ (b) $0.65 = \frac{65}{100} = \frac{13}{20}$ (c) $0.07 = \frac{7}{100}$

- Be able to carry out a variety of calculations involving decimals.
- Know that when a number is:
multiplied by a number between 0 and 1 the result will be **smaller** than the original number,
divided by a number between 0 and 1 the result will be **larger** than the original number.

Exercise 2

Do not use a calculator for questions 1 to 13.

1. Look at this collection of numbers.
Two of these numbers are multiplied together.
Which two numbers will give the smallest answer?

13.5 0.065 0.9 23.0 4.5

2 Write the decimals 1.18, 1.80, 1.08, 1.118 in order, smallest first.

3 Work out.
(a) £2.15 + £20 (b) 6.74×100 (c) 4.5×2 (d) $64.8 \div 4$ OCR

4 Toyah buys the following vegetables.
0.55 kg onions 1.2 kg carrots 2.5 kg potatoes 0.65 kg leeks
What is the total weight of the vegetables?

5 (a) John buys 3 pens at 40 pence each.
How much change should he get from a £5 note?
(b) Julie has £15 and wants to buy as many magazines as possible. Each magazine cost £2.99.
How many can she buy? OCR

6 Nick hires a digger.
Calculate the total cost of hiring the digger for 6 hours.

DIGGER FOR HIRE
Fixed charge £16.50
plus £20.20 per hour

OCR

7 Mrs Jolly buys 5 graphical calculators. They cost £137.25 altogether.
Work out the cost of one graphical calculator. OCR

8 Beverley buys crisps and cola.
Here is the bill.
The corner has been torn off.
Work out the cost of one can of cola.

5 cans of cola	
5 packets of crisps	£1.55
TOTAL	£3.45

OCR

9 Calculate the total cost of 16 DVDs at £14.32 each. OCR

10 Lucy works out 0.2×0.4. She gets the answer 0.8.
Explain why her answer must be wrong.

11 Work out. (a) 0.8×0.2 (b) $31.2 \div 4$ OCR

12 Using the calculation $23 \times 32 = 736$, work out the following.
(a) 2.3×3.2 (b) $73.6 \div 23$ (c) $736 \div 3.2$

13 Write as a fraction. (a) 0.3 (b) 0.03 (c) 0.33

14 Kevin is working out the time needed to complete a journey.
Using his calculator, he gets the answer 0.66666666
The result is in hours.
How many minutes will the journey take?

15 $5 \times m$ gives an answer **less than 5**. $5 \div m$ gives an answer **more than 5**.
Give two possible values for m which satisfy **both** conditions.

16 Potatoes are sold in bags and sacks.
Bags of potatoes weigh 2.5 kg and cost 95 pence.
Sacks of potatoes weigh 12 kg and cost £3.18.
How much, per kilogram, is saved by buying sacks of potatoes instead of bags of potatoes?

17 Wayne buys 2 kg of apples and 0.5 kg of cherries.
The total cost is £2.85. The apples cost 80p per kilogram.
How much per kilogram do cherries cost? OCR

18 Work out $\frac{12.9 \times 7.3}{3.9 + 1.4}$. Write down your full calculator display.

SECTION 3

Approximation and Estimation

What you need to know

- How to **round** to the nearest 10, 100, 1000.

Eg 1 Write 6473 to (a) the nearest 10, (b) the nearest 100, (c) the nearest 1000.
(a) 6470, (b) 6500, (c) 6000.

- In real-life problems a rounding must be used which gives a sensible answer.

Eg 2 Doughnuts are sold in packets of 6. Tessa needs 20 doughnuts for a party.
How many packets of doughnuts must she buy?

$20 \div 6 = 3.33\ldots$ This should be rounded up to 4. So, Tessa must buy 4 packets.

- How to approximate using **decimal places**.

> Write the number using one more decimal place than asked for.
> Look at the last decimal place and
> - if the figure is 5 or more round up,
> - if the figure is less than 5 round down.

Eg 3 Write the number 3.649 to
(a) 2 decimal places,
(b) 1 decimal place.

(a) 3.65,
(b) 3.6.

- How to approximate using **significant figures**.

> Start from the most significant figure and count the required number of figures.
> Look at the next figure to the right of this and
> - if the figure is 5 or more round up,
> - if the figure is less than 5 round down.
>
> Add noughts, as necessary, to preserve the place value.

Eg 4 Write each of these numbers correct to 2 significant figures.
(a) 365
(b) 0.0423

(a) 370
(b) 0.042

- You should be able to choose a suitable degree of accuracy.

> The result of a calculation involving measurement should not be given to a greater degree of accuracy than the measurements used in the calculation.

- Be able to use approximations to estimate that the actual answer to a calculation is of the right order of magnitude.

> Estimation is done by approximating every number in the calculation to one significant figure.
> The calculation is then done using the approximated values.

Eg 5 Use approximations to estimate $\frac{5.1 \times 57.2}{9.8}$.

$$\frac{5.1 \times 57.2}{9.8} = \frac{5 \times 60}{10} = 30$$

- Be able to use a calculator to check answers to calculations.
- Be able to recognise limitations on the accuracy of data and measurements.

Eg 6 Jamie said, "I have 60 friends at my party." This figure is correct to the nearest 10.
What is the smallest and largest possible number of friends Jamie had at his party?

The smallest whole number that rounds to 60 is 55.
The largest whole number that rounds to 60 is 64.
So, smallest is 55 friends and largest is 64 friends.

Eg 7 A man weighs 57 kg, correct to the nearest kilogram.
What is the minimum weight of the man?
Minimum weight = 57 kg − 0.5 kg = 56.5 kg.

Exercise 3

Do not use a calculator for questions 1 to 21.

1 Write the result shown on the calculator display
(a) to the nearest whole number,
(b) to the nearest ten,
(c) to the nearest hundred.

626.47

2 (a) Write 478 correct to the nearest ten.
(b) Write 4290 correct to the nearest thousand. OCR

3 A newspaper's headline states: "20 000 people attend concert".
The number in the newspaper is given to the nearest thousand.
What is the smallest possible attendance?

4 The diagram shows the distances between towns A, B and C.

By rounding each of the distances given to the nearest hundred, estimate the distance between A and C.

5 Wayne is calculating $\frac{8961}{1315 + 1692}$.
(a) Write down each of the numbers 8961, 1315 and 1692 to the nearest hundred.
(b) Hence, estimate the value of $\frac{8961}{1315 + 1692}$.

6 On Saturday a dairy sold 2975 litres of milk at 42 pence per litre.
By rounding each number to one significant figure, estimate the amount of money received from the sale of milk, giving your answer in pounds.

7

(a) Use approximation to show that this is correct.
(b) (i) Show how you could find an estimate for $2019 \div 37$.
(ii) What is your estimated answer?

8 (a) Work out $75.6 \div 27$.
(b) What approximate calculation could you do to check the answer to part (a)? OCR

9 Mary earned £21 788 in 52 weeks.
Estimate how much she earned per week, showing how you found it. OCR

10 (a) To estimate 97×49, Charlie uses the approximations 100×50.
Explain why his estimate will be larger than the actual answer.
(b) To estimate $1067 \div 48$, Patsy uses the approximations $1000 \div 50$.
Will her estimate be larger or smaller than the actual answer?
Give a reason for your answer.

11 Jim has done this calculation. $58\,900 \div 62 = 95$. His answer is wrong.
Explain how you can tell the answer is wrong without working it out exactly. OCR

12 The tickets for a concert cost £12.80 each.
(a) A group of 24 people goes to the concert.
Calculate the total cost of their tickets.
(b) The total number of tickets sold for the concert was 20 287.
Estimate the total money paid for these tickets. OCR

13 The length of a garden is 50 m, correct to the nearest metre.
What is the minimum length of the garden?

14 A car park has spaces for 640 cars, correct to the nearest ten.
(a) What is the least possible number of spaces in the car park?
(b) What is the greatest possible number of spaces in the car park?

15 Mrs Patel is buying some history books. The books cost £6.95 each.
She wants to estimate the cost of 39 books.
(a) Write down a calculation she could do in her head to work out an estimate for the total cost.
(b) Is her estimate bigger or smaller than the exact cost?
Explain how you decided. OCR

16 Melanie needs 200 crackers for an office party.
The crackers are sold in boxes of 12.
How many boxes must she buy?

17 Clint has to calculate $\frac{414 + 198}{36}$. He calculates the answer to be 419.5.

By rounding each number to one significant figure, estimate whether his answer is about right.
Show all your working.

18 (a) A group of 17 people win £59 372 in a lottery.
They share the money equally between them.
Estimate how much money they will each receive.
Show how you worked out your estimate.

(b) Work out an **estimate** for the value of $\frac{51 \times 38}{0.47}$.
Show how you worked out your estimate. OCR

19 (a) Find an approximate value of $\frac{21 \times 58}{112}$.

(b) Use a calculator to find the difference between your approximate value and the exact value.

20 In 2005, Mr Symms drove 8873 kilometres.
His car does 11 kilometres per litre. Petrol costs 89.9 pence per litre.
(a) By rounding each number to one significant figure, estimate the amount he spent on petrol.
(b) Without any further calculation, explain why this estimate will be larger than the actual amount.

21 Georgina said, "I spent £100 on my holidays." This amount is given correct to the nearest £10.
Write down the minimum and maximum amounts Georgina could have spent.

22 Calculate $97.2 \div 6.5$.
Give your answer correct to (a) two decimal places, (b) one decimal place.

23 Calculate 78.4×8.7.
Give your answer correct to (a) two significant figures, (b) one significant figure.

24 Andrew says, "Answers given to two decimal places are more accurate than answers given to two significant figures." Is he right? Explain your answer.

25 Calculate the value of $\frac{65.4}{4.3 + 3.58}$.

(a) Write down your full calculator display.
(b) Give your answer correct to 3 significant figures.

26 Calculate. $\frac{6.5 \times 4.7}{6.7 - 1.9}$ Give your answer correct to 1 decimal place. OCR

27 Use your calculator to evaluate the following. $\frac{50 - 19.7}{31.6 + 55.1}$
Give your answer correct to one decimal place.

Negative Numbers

What you need to know

- You should be able to use **negative numbers** in context, such as temperature, bank accounts.
- Realise where negative numbers come on a **number line**.

$-5 \quad -4 \quad -3 \quad -2 \quad -1 \quad 0 \quad 1 \quad 2 \quad 3 \quad 4 \quad 5 \quad 6$

- Be able to put numbers in order (including negative numbers).

Eg 1 Write the numbers 19, -3, 7, -5 and 0 in order, starting with the smallest.
-5, -3, 0, 7, 19

- You should be able to add, subtract, multiply and divide with negative numbers.

Eg 2 Work out.
(a) $-3 + 10$
$= 7$
(b) $-5 - 7$
$= -12$
(c) -4×5
$= -20$
(d) $-12 \div 4$
$= -3$

- Be able to use these rules with negative numbers.

When adding or subtracting:
+ + can be replaced by +
− − can be replaced by +
+ − can be replaced by −
− + can be replaced by −

When multiplying:
$+ \times + = +$
$- \times - = +$
$+ \times - = -$
$- \times + = -$

When dividing:
$+ \div + = +$
$- \div - = +$
$+ \div - = -$
$- \div + = -$

Eg 3 Work out.
(a) $(-3) + (-2)$
$= -3 - 2$
$= -5$
(b) $(-5) - (-8)$
$= -5 + 8$
$= 3$
(c) $(-2) \times (-3)$
$= 6$
(d) $(-8) \div (+2)$
$= -4$

Exercise 4

Do not use a calculator for this exercise.

1. What temperatures are shown by these thermometers?

(a)

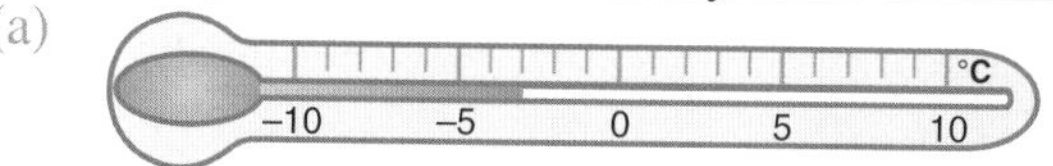

(b)

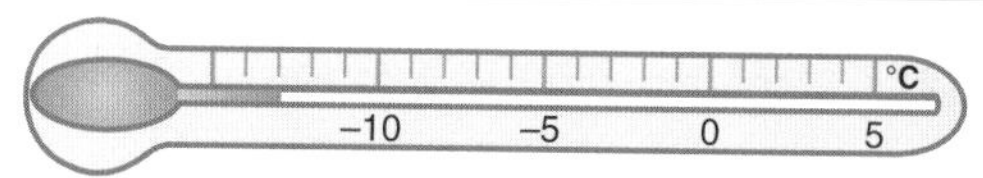

2. The midday temperatures in three different places on the same day are shown.

Moscow −7°C	Oslo −9°C	Warsaw −5°C

(a) Which place was coldest?
(b) Which place was warmest?

3. The top of a cliff is 125 m above sea level.
The bottom of a lake is 15 m below sea level.
How far is the bottom of the lake below the top of the cliff?

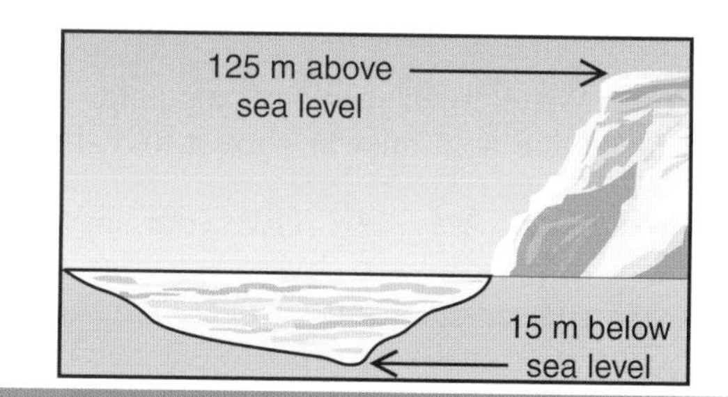

4 Place the following numbers in order of size, starting with the smallest.

17 −9 −3 5 0 7

5 The temperatures, in °C, in five cities are listed in the table.

City	Amsterdam	Dublin	Oslo	Venice	Warsaw
Temperature (°C)	−4	5	−11	6	−9

(a) (i) Which city is the warmest?
(ii) Which city is the coldest?
(b) The temperature in Budapest is 2 degrees colder than Amsterdam.
What is the temperature in Budapest? OCR

6 Gordon has £28 in his bank account.
He pays a bill of £85 by cheque, which is accepted by his bank.
What is the new balance in his account?

7 The table shows the temperatures recorded at a ski resort one day in February.

Time	0600	1200	1800	2400
Temperature (°C)	−3	3	−2	−6

(a) By how many degrees did the temperature rise between 0600 and 1200?
(b) During which six-hourly period was the maximum drop in temperature recorded?

8 Find the missing numbers so that each row adds up to 5.

(a) | −2 | 3 | | (b) | −2 | | −4 | (c) | | 9 | −1 |

9 What number must be placed in the box to complete each of the following?

(a) $-4 + \square = 2$ (b) $-2 - \square = -7$ (c) $-2 - \square = -1$ (d) $\square \times 4 = -12$

10 The ice cream is stored at −25°C.
How many degrees is this below the required storage temperature?

11 At Death Valley in California, Bad Water is 86 metres below sea level and Telescope Peak is 3368 metres above sea level.
(a) Calculate the difference in height between Bad Water and Telescope Peak.
(b) As you go up from Bad Water, the temperature drops by 1°C every 80 metres.
The temperature at Bad Water is 45°C.
Calculate the temperature at Telescope Peak.
Give your answer to a suitable degree of accuracy. OCR

12 Complete these calculations.

(a) $-7 + 12 = \ldots$ (b) $-3 \times 10 = \ldots$ (c) $\ldots - 6 = -8$ (d) $-20 \div \ldots = 4$

OCR

13 This rule can be used to estimate the temperature in °F for temperatures given in °C.

Multiply the temperature in °C by 2 and add 30.

Use this rule to estimate −5°C in °F.

14 Work out. (a) $\dfrac{(-2) \times (-5) \times (+6)}{(-3)}$ (b) $(-3) + (-2) \times (+6)$

15 A test has 12 questions.

A correct answer scores +3 marks. An incorrect answer scores −1 mark.

Pippa attempts every question and scores 8 marks.
How many correct answers did she get?

SECTION 5

Fractions

What you need to know

- The top number of a fraction is called the **numerator**, the bottom number is called the **denominator**.
- Fractions which are equal are called **equivalent fractions**.

To write an equivalent fraction:
Multiply the numerator and denominator by the **same** number.

Eg 1 $\frac{1}{4} = \frac{1 \times 3}{4 \times 3} = \frac{1 \times 5}{4 \times 5}$

$\frac{1}{4} = \frac{3}{12} = \frac{5}{20}$

- Fractions can be **simplified** if both the numerator and denominator can be divided by the **same number**. This is sometimes called **cancelling**.

Eg 2 Write $\frac{20}{28}$ as a fraction in its simplest form.

$\frac{20}{28} = \frac{20 \div 4}{28 \div 4} = \frac{5}{7}$

Divide the numerator and denominator by the largest number that divides into both.

- $2\frac{1}{2}$ is an example of a **mixed number**.
  It is a mixture of whole numbers and fractions.
- $\frac{5}{2}$ is an **improper** (or '**top heavy**') fraction.
- Fractions must have the **same denominator** before **adding** or **subtracting**.

Eg 3 Work out.

(a) $\frac{4}{5} - \frac{1}{2} = \frac{8}{10} - \frac{5}{10} = \frac{3}{10}$

(b) $2\frac{3}{4} + 1\frac{2}{3} = 2\frac{9}{12} + 1\frac{8}{12} = 3\frac{17}{12} = 4\frac{5}{12}$

Add (or subtract) the numerators only. When the answer is an improper fraction change it into a mixed number.

- You should be able to multiply and divide fractions.

Eg 4 Work out.

(a) $\frac{3}{4} \times \frac{2}{3} = \frac{\not{3}^{1}}{\not{4}_{2}} \times \frac{\not{2}^{1}}{\not{3}_{1}} = \frac{1}{2}$

The working can be simplified by cancelling.

(b) $\frac{3}{4} \div \frac{2}{3} = \frac{3}{4} \times \frac{3}{2} = \frac{9}{8} = 1\frac{1}{8}$

Dividing by $\frac{2}{3}$ is the same as multiplying by $\frac{3}{2}$.

- All fractions can be written as decimals.

To change a fraction to a decimal divide the **numerator** by the **denominator**.

Eg 5 Change $\frac{4}{5}$ to a decimal.

$\frac{4}{5} = 4 \div 5 = 0.8$

- Some decimals have **recurring digits**.
  These are shown by:
  a single dot above a single recurring digit,

  Eg 6 $\frac{2}{3} = 0.6666\ldots = 0.\dot{6}$

  a dot above the first and last digit of a set of recurring digits.

  Eg 7 $\frac{5}{11} = 0.454545\ldots = 0.\dot{4}\dot{5}$

Exercise 5

Do not use a calculator for this exercise.

1 (a) What fraction of this rectangle is shaded?

(b) Copy and shade $\frac{2}{3}$ of this rectangle.

2 Work out $\frac{3}{4}$ of £32.

3 (a) Work out $\frac{1}{4}$ of 28.
(b) Write $\frac{3}{4}$ as a decimal.
(c) Which of the following fractions are equal to $\frac{1}{3}$? $\frac{3}{5}$ $\frac{2}{6}$ $\frac{6}{12}$ $\frac{10}{30}$ OCR

4 (a) Find $\frac{5}{8}$ of £48.
(b) Work out. $\frac{5}{8} \times \frac{1}{2}$
(c) Write these fractions in order of size, starting with the smallest.

$\frac{2}{3}$ $\frac{5}{12}$ $\frac{11}{24}$ $\frac{1}{2}$ $\frac{5}{8}$ OCR

5 (a) Which of the fractions $\frac{7}{10}$ or $\frac{4}{5}$ is the smaller? Explain why.
(b) Write down a fraction that lies halfway between $\frac{1}{3}$ and $\frac{1}{2}$.

6 (a) Write this fraction in its simplest form. $\frac{35}{45}$
(b) Change $\frac{3}{8}$ into a decimal.
(c) Work out. $\frac{1}{4} + \frac{5}{8}$ OCR

7 Jan uses $\frac{3}{4}$ of a jar of cherries to make a cheesecake.
How many jars of cherries does she need to buy to make 10 cheesecakes?

8 An examination is marked out of 48. Ashley scored 32 marks.
What fraction of the total did he score?
Give your answer in its simplest form.

9 Calculate (a) $\frac{2}{9} \times 3$, (b) $\frac{6}{7} \div 4$. Give your answers in their simplest form. OCR

10 The cake stall at a school fete has 200 fairy cakes for sale.
It sells $\frac{3}{5}$ of them at 25p each and the remainder at 20p each.
How much money does the stall get from selling fairy cakes?

11 George buys $\frac{1}{4}$ kg of jellies at £3.60 per kilogram and $\frac{1}{5}$ kg of toffees at £4.80 per kilogram.
How much change does he get from £5?

12 Work out. (a) $\frac{2}{5} \times \frac{1}{4}$ (b) $\frac{2}{5} - \frac{1}{4}$ OCR

13 (a) Change $\frac{1}{6}$ to a decimal. Give the answer correct to 3 d.p.
(b) Write these numbers in order of size, starting with the largest.

1.067 1.7 1.66 $1\frac{1}{6}$ 1.67

14 Income tax and national insurance take $\frac{1}{5}$ of Phillip's pay.
He gives $\frac{2}{5}$ of what he has left to his parents for housekeeping.
What fraction of his pay does Phillip have left for himself?

15 Three-fifths of the people at a party are boys.
Three-quarters of the boys are wearing fancy dress.
What fraction of the people at the party are boys wearing fancy dress?

16 Calculate. (a) $1\frac{2}{3} - \frac{1}{4}$ (b) $\frac{3}{5} \div \frac{2}{3}$ OCR

17 Stuart pays £3.50 for $\frac{1}{4}$ kg of Stilton Cheese and $\frac{1}{2}$ kg of Cheddar Cheese.
Stilton Cheese costs £6.40 per kilogram. How much per kilogram is Cheddar Cheese?

18 (a) Use your calculator to change $\frac{3}{16}$ into an exact decimal.
(b) From the list of fractions, choose the fraction that is nearest to 0.5.

$\frac{3}{7}$ $\frac{5}{9}$ $\frac{2}{5}$ $\frac{6}{11}$

Show clear working to support your answer. OCR

Working with Number

What you need to know

- **Multiples** of a number are found by multiplying the number by 1, 2, 3, 4, ...

  Eg 1 The multiples of 8 are $1 \times 8 = \mathbf{8}$, $2 \times 8 = \mathbf{16}$, $3 \times 8 = \mathbf{24}$, $4 \times 8 = \mathbf{32}$, ...

- **Factors** of a number are found by listing all the products that give the number.

  Eg 2 $1 \times 6 = 6$ and $2 \times 3 = 6$.
  So, the factors of 6 are: 1, 2, 3 and 6.

- The **common factors** of two numbers are the numbers which are factors of **both**.

  Eg 3 Factors of 16 are: 1, 2, 4, 8, 16.
  Factors of 24 are: 1, 2, 3, 4, 6, 8, 12, 24.
  Common factors of 16 and 24 are: 1, 2, 4, 8.

- A **prime number** is a number with only two factors, 1 and the number itself.
  The first few prime numbers are: 2, 3, 5, 7, 11, 13, 17, 19, ...
  The number 1 is not a prime number because it has only one factor.

- The **prime factors** of a number are those factors of the number which are prime numbers.

  Eg 4 The factors of 18 are: 1, 2, 3, 6, 9 and 18.
  The prime factors of 18 are: 2 and 3.

- The **Least Common Multiple** of two numbers is the smallest number that is a multiple of both.

  Eg 5 The Least Common Multiple of 4 and 5 is 20.

- The **Highest Common Factor** of two numbers is the largest number that is a factor of both.

  Eg 6 The Highest Common Factor of 8 and 12 is 4.

- An expression such as $3 \times 3 \times 3 \times 3 \times 3$ can be written in a shorthand way as 3^5.
  This is read as '3 to the power of 5'.
  The number 3 is the **base** of the expression. 5 is the **power**.

- Powers can be used to help write any number as the **product of its prime factors**.

  Eg 7 $72 = 2 \times 2 \times 2 \times 3 \times 3 = 2^3 \times 3^2$

- Numbers raised to the power of 2 are **squared**.
  For example, $3^2 = 3 \times 3 = 9$.
  Squares can be calculated using the x^2 button on a calculator.

  > **Square numbers** are whole numbers squared.
  > The first few square numbers are: 1, 4, 9, 16, 25, 36, ...

  The opposite of squaring a number is called finding the **square root**.
  Square roots can be found by using the $\sqrt{}$ button on a calculator.
  The square root of a number can be positive or negative.

  Eg 8 The square root of 9 is $+3$ or -3.

- You should be able to find square roots using a method called **trial and improvement**.
  Work methodically using trials first to the nearest whole number, then to one decimal place, etc.
  Do one trial to one more decimal place than the required accuracy to be sure of your answer.

- Numbers raised to the power of 3 are **cubed**.
  For example, $4^3 = 4 \times 4 \times 4 = 64$.

  Cube numbers are whole numbers cubed.
  The first few cube numbers are: 1, 8, 27, 64, 125, ...

  The opposite of cubing a number is called finding the **cube root**.

  Cube roots can be found by using the [$\sqrt[3]{\ }$] button on a calculator.

- **Powers**
  The squares and cubes of numbers can be worked out on a calculator by using the [x^y] button.

  The [x^y] button can be used to calculate the value of a number x raised to the power of y.

  Eg 9 Calculate 2.6^4.
  Enter the sequence: [2] [.] [6] [x^y] [4] [=]. So, $2.6^4 = 45.6976$.

- The **reciprocal** of a number is the value obtained when the number is divided into 1.

  Eg 10 The reciprocal of 2 is $\frac{1}{2}$.

  The reciprocal of a number can be found on a calculator by using the [$\frac{1}{x}$] button.
  A number times its reciprocal equals 1. Zero has no reciprocal.

- You should be able to simplify calculations involving powers.
  Powers of the same base are **added** when terms are **multiplied**.
  Powers of the same base are **subtracted** when terms are **divided**.

  In general:
  $a^m \times a^n = a^{m+n}$
  $a^m \div a^n = a^{m-n}$

  Eg 11 (a) $2^3 \times 2^2 = 2^5$ (b) $2^3 \div 2^2 = 2^1 = 2$

You should be able to:

- use the [x^2], [x^y], [$\sqrt{\ }$] and [$\frac{1}{x}$] buttons on a calculator to solve a variety of problems.
- interpret a calculator display for very large and very small numbers expressed in standard index form.

  Eg 12 [1.5   10] means $1.5 \times 10^{10} = 15\,000\,000\,000$

  [6.2   −05] means $6.2 \times 10^{-5} = 0.000\,062$

Exercise 6

Do not use a calculator for questions 1 to 19.

1 Choose one number from this list to complete each sentence.

3   4   5   7   16   18   21   32   49

(a) .... is a factor of 8.
(b) .... is a multiple of 6.
(c) .... is a common factor of 14 and 35. OCR

2 (a) Write down all the factors of 18.
(b) Write down a multiple of 7 between 30 and 40.
(c) Explain why 9 is not a prime number.
(d) Find the common factors of 18 and 24.

3 (a) What is the number?
It is • bigger than 10 • less than 20 • a square number.
(b) What are the numbers?
They are • bigger than 10 • less than 30 • common factors of 48 and 72.
(c) What are the numbers?
They are • bigger than 20 • less than 100 • cube numbers. OCR

4 (a) What is the square root of 100?
(b) What is the cube root of 8?

5 Write down the values of the following. (a) 7^2 (b) $\sqrt{64}$ OCR

6 Jenny says that $2^2 + 3^2 = (2 + 3)^2$. Is she right? Show your working.

7 Work out the following.
(a) the cube of 3 (b) 2^4 (c) 0.3^2 (d) an **estimate** of 8.8^2 OCR

8 Find the value of (a) $1^2 + 2^2 + 3^2 + 4^2 + 5^2$, (b) $9^2 \times 10^2$, (c) $2^3 \times 5^2$.

9 (a) Work out. (i) 10^3 (ii) $\frac{2^5}{4^2}$ (iii) 0.6^2
(b) 21 22 23 24 25 26 27 28 29
From these numbers, choose one which is:
(i) a cube number, (ii) a prime number. OCR

10 Work out. (a) $2^3 \times 3^2$ (b) $\sqrt{25} + \sqrt{144}$ (c) $\sqrt{49} \times 4^2$

11 (a) Simplify. $\frac{7^2 \times 7^3}{7^6}$
(b) Write down $\sqrt{169}$.
(c) Write 120 as a product of its prime factors. OCR

12 Find the highest common factor (HCF) of 12 and 20. OCR

13 A white light flashes every 10 seconds. A red light flashes every 6 seconds.
The two lights flash at the same time.
After how many seconds will the lights next flash at the same time?

14 (a) What is the cube root of 125?
(b) What is the reciprocal of 4?
(c) Which is smaller $\sqrt{225}$ or 2^4? Show your working.

15 Find the value of x in each of the following.
(a) $7^6 \times 7^3 = 7^x$ (b) $7^6 \div 7^3 = 7^x$

16 Simplify fully each of these expressions. Leave your answers in power form.
(a) $3^2 \times 3^3$ (b) $5^6 \div 5^3$ (c) $\frac{2 \times 2^3}{2^2}$

17 (a) Between which two consecutive whole numbers does $\sqrt{70}$ lie?
(b) Use a trial and improvement method to find the square root of 70, correct to two decimal places. Show your working clearly.

18 (a) Find the reciprocal of 7, correct to two decimal places.
(b) Find the value of 5.6^3.

19 Calculate $\frac{50 + \sqrt{12}}{6.8}$. Give your answer correct to two decimal places. OCR

20 Calculate $\frac{3.82^2}{3.41 - 1.25}$. Give your answer correct to 2 decimal places. OCR

21 Calculate. (a) $\frac{2.45 + 1.474}{4.25 - 3.53}$ (b) $\sqrt{2.43^2 + 1.65^2}$ OCR

22 Calculate. $2.5^3 - 1.6^2 \times 4.75$ OCR

23 (a) Calculate the value of $\sqrt{\frac{4.1}{(0.19)^2}}$.
(b) Show how to check that your answer is of the right order of magnitude.

SECTION 

Percentages

What you need to know

- 10% is read as '10 percent'. 'Per cent' means out of 100. 10% means 10 out of 100.
- A percentage can be written as a fraction, 10% can be written as $\frac{10}{100}$.
- To change a decimal or a fraction to a percentage: **multiply by 100**.

Eg 1 Write as a percentage (a) 0.12 (b) $\frac{8}{25}$

(a) $0.12 \times 100 = 12\%$ (b) $\frac{8}{25} \times 100 = 32\%$

- To change a percentage to a fraction or a decimal: **divide by 100**.

Eg 2 Write 18% as (a) a decimal, (b) a fraction.

(a) $18\% = 18 \div 100 = 0.18$, (b) $18\% = \frac{18}{100} = \frac{9}{50}$.

- How to express one quantity as a percentage of another.

Eg 3 Write 30p as a percentage of £2.

$\frac{30}{200} \times 100 = 30 \times 100 \div 200 = 15\%$

> Write the numbers as a fraction, using the same units.
> Change the fraction to a percentage.

- You should be able to use percentages to solve a variety of problems.
- Be able to find a percentage of a quantity.

Eg 4 Find 20% of £64.

£64 ÷ 100 = £0.64

£0.64 × 20 = £12.80

> 1. Divide by 100 to find 1%.
> 2. Multiply by the percentage to be found.

- Be able to find a percentage increase (or decrease).

$$\text{Percentage increase} = \frac{\text{actual increase}}{\text{initial value}} \times 100\%$$

$$\text{Percentage decrease} = \frac{\text{actual decrease}}{\text{initial value}} \times 100\%$$

Eg 5 Find the percentage loss on a micro-scooter bought for £25 and sold for £18.

Percentage loss $= \frac{7}{25} \times 100 = 28\%$

Exercise 7

Do not use a calculator for questions 1 to 12.

1. What percentage of these rectangles are shaded?

(a)

(b) 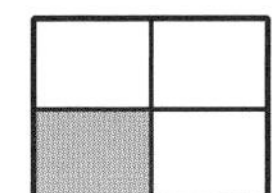

(c) 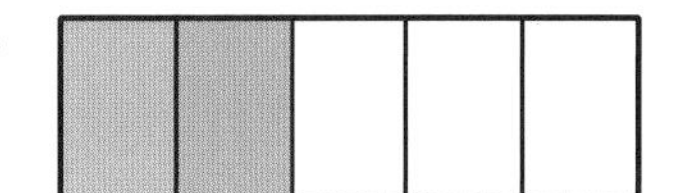

2. Write $\frac{1}{2}$, 0.02 and 20% in order of size, smallest first.

3. Work out (a) 10% of 20 pence, (b) 25% of 60 kg, (c) 5% of £900.

4. In an examination, Felicity scored 75% of the marks and Daisy scored $\frac{4}{5}$ of the marks.
Who has the better score? Give a reason for your answer.

5. Work out 30% of £45.

6. In Year 7 there are 110 boys and 80 girls.
10% of the boys wear glasses and 15% of the girls wear glasses.
How many more girls than boys wear glasses? OCR

7. A pop concert is attended by 35 000 people. 2% of the people are given a free T-shirt.
How many people are given a free T-shirt?

8. Andy is given £8 pocket money. He spends 15% of it on a magazine.
How much was the magazine?

9. Mira earns £600 a week. She is given a pay rise of £30 a week.
What is the percentage increase in her pay?

10. 180 college students apply for jobs at a new supermarket.
(a) 70% of the students are given an interview.
How many students are given an interview?
(b) 54 students are offered jobs.
What percentage of the students who applied were offered jobs?

11. A mobile phone normally costs £90. The price is reduced by 20% in a sale.
What is the price of the mobile phone in the sale?

12. Maggie normally works Monday to Friday and is paid £6.50 per hour.
When she works on a Saturday she is paid 30% **more**.
How much is she paid per hour for working on a Saturday?

13. Rosie scored 72% in a test.
Francine scored 52 out of 75 marks in the same test.
Who did better in the test? Give a reason for your answer. OCR

14. In 2005, Manor Park had 84 400 visitors.
In 2006, there was an 8% increase in the number of visitors.
How many people visited Manor Park in 2006? OCR

15. In an experiment a spring is extended from 12 cm to 15 cm.
Calculate the percentage increase in the length of the spring.

16. A pogo stick is bought for £12.50 and sold for £8.
What is the percentage loss?

17. Bob, Carl and Peter share a flat. The rent for the flat is £725.
Bob pays 42% of the rent. Carl pays $\frac{3}{10}$ of the rent. Peter pays the rest.
How much does Peter pay? OCR

18. You have to climb 123 steps to see the view from the top of a tower.
Harold has climbed 66 steps.
What percentage of the steps has he still got to climb?
Give your answer to the nearest whole number.

19. A farmer has 200 sheep. 90% of the sheep have lambs.
Of the sheep which have lambs 45% have two lambs.
How many of the sheep have two lambs?

20. In a sale, all items are reduced by 20%.
If a sale item is not sold by the end of the week, the sale price is reduced by a further 20%.
A bed originally costs £300.
It is not sold by the end of the sale week and so is subject to the further reduction.
(a) Calculate the final price of the bed.
(b) Calculate the overall percentage reduction on the original price. OCR

Time and Money

What you need to know

- Time can be given using either the **12-hour clock** or the **24-hour clock**.

Eg 1 (a) 1120 is equivalent to 11.20 am.
(b) 1645 is equivalent to 4.45 pm.

When using the 12-hour clock:
times **before** midday are given as am,
times **after** midday are given as pm.

- **Timetables** are usually given using the 24-hour clock.

Eg 2 Some of the rail services from Manchester to Stoke are shown.

Manchester	0925	1115	1215	1415	1555
Stockport	0933	—	1223	—	1603
Stoke	1007	1155	1255	1459	1636

Some trains do not stop at every station. This is shown by a dash on the timetable.

Kath catches the 1555 from Manchester to Stoke.
(a) How many minutes does the journey take? (a) 41 minutes.
(b) What is her arrival time in 12-hour clock time? (b) 4.36 pm.

- When considering a **best buy**, compare quantities by using the same units.

Eg 3 Peanut butter is available in small or large jars.
Small jar: 250 grams for 68 pence. Large jar: 454 grams for £1.25.
Which size is the better value for money?

Compare the number of grams per penny for each size.

Small jar: 250 ÷ 68 = 3.67… grams per penny.
Large jar: 454 ÷ 125 = 3.63… grams per penny.
The small jar gives more grams per penny and is better value.

- **Value added tax**, or **VAT**, is a tax on some goods and services and is added to the bill.

Eg 4 A freezer costs £180 + $17\frac{1}{2}$% VAT.

$17\frac{1}{2}\% = 17.5\% = \frac{17.5}{100} = 0.175$

(a) How much is the VAT? (a) VAT = £180 × 0.175 = £31.50
(b) What is the total cost of the freezer? (b) Total cost = £180 + £31.50 = £211.50

- **Exchange rates** are used to show what £1 will buy in foreign currencies.

Eg 5 Alex buys a painting for 80 euros in France.
The exchange rate is 1.55 euros to the £.
What is the cost of the painting in £s?

1.55 euros = £1 80 euros = 80 ÷ 1.55 = £51.6129…
The painting cost £51.61, to the nearest penny.

Exercise 8

Do not use a calculator for questions 1 to 6.

1. Nick is on holiday in Spain.
He hires a car at the rates shown.

There are 1.60 euros to £1.

Nick hires the car for 5 days and drives it for a total of 720 kilometres.
Calculate the total cost of hiring the car.
Give your answer in pounds.

CAR HIRE

Daily rate	54 euros
Free kilometres per day	120
Excess kilometre charge	0.60 euros

2 This is part of the timetable for the Ullswater ferry service.

Glenridding	1000	1100	1130	1300	1400	1440	1600	1630
Howtown	1035	1135	1205	1335	1435	1515	1635	1705
Pooley Bridge	—	1155	—	—	1455	—	—	1725

(a) Sam gets the ferry at Glenridding at 1400.
At what time does he get to Pooley Bridge?

(b) Mr and Mrs Wilson get the 1630 ferry from Glenridding to Howtown.
(i) How many minutes does the journey take?
(ii) A one-way adult ticket from Glenridding to Howtown is £2.60.
Mr and Mrs Wilson travel from Glenridding to Howtown and back.
Work out the total cost for their ferry journeys. OCR

3 Two supermarkets have special offers on cans of Kola.

Which supermarket gives the better deal when you buy 5 cans?
Show how you got your answer.

Amart
Normal price 45p a can.
Special offer
5 cans for the price of 4.

Bazda
Normal price 47p a can.
Special offer
50p off when you buy 5 cans.

OCR

4 The table below shows the cost of hiring a wallpaper stripper.

Cost for the first day	Extra cost per day for each additional day
£7.50	£2.50

Vivian hires the wallpaper stripper. The total cost of hiring the wallpaper stripper was £35.
How many days did Vivian hire it for?

5 The cash price of a washing machine is £450.
Instead of paying cash, James pays a deposit of £50 and then makes twelve monthly payments of £38.50 each. How much more will James pay by this method than by paying cash? OCR

6 A market trader buys 25 kg of apples at 40p per kg.
She makes 70% profit on the first 20 kg of apples that she sells.
The rest she sells at only $\frac{4}{5}$ of what she paid for them. How much profit does she make? OCR

7 Toffee is sold in bars of two sizes.
A large bar weighs 450 g and costs £1.69. A small bar weighs 275 g and costs 99p.
Which size of bar is better value for money? You must show all your working.

8 Mrs Tilsed wishes to buy a car priced at £2400.

Two options are available.
Option 1 – A deposit of 20% of £2400 and 24 monthly payments of £95.
Option 2 – For a single payment the dealer offers a discount of 5% on £2400.

How much more does it cost to buy the car if option 1 is chosen rather than option 2?

9 A French supermarket buys coffee for 3.90 euros per kilogram.
(a) The supermarket sells the coffee to make a profit of 60%.
Calculate the selling price of one kilogram of coffee.
(b) A British importer also buys the coffee at 3.90 euros per kilogram.
The exchange rate is £1 = 1.45 euros.
Calculate the cost of one kilogram of coffee in British money.
Give your answer to an appropriate degree of accuracy. OCR

10 Paul wants to buy a new computer.
At *PC Essentials*, Paul needs to pay £890 plus VAT at $17\frac{1}{2}$%.
At *Computers For All*, the total price is £999.
Find the difference in the price of the computers. OCR

SECTION 9 Personal Finance

What you need to know

- **Hourly pay** is paid at a **basic rate** for a fixed number of hours.
  Overtime pay is usually paid at a higher rate such as time and a half, which means each hour's work is worth 1.5 times the basic rate.

  Eg 1 Alexis is paid £7.20 per hour for a basic 35-hour week.
  Overtime is paid at time and a half.
  Last week she worked 38 hours. How much was Alexis paid last week?

  Basic pay = $£7.20 \times 35$ = £252
  Overtime pay = $1.5 \times £7.20 \times 3$ = £ 32.40
  Total pay = $£252 + £32.40$ = £284.40

- Everyone is allowed to earn some money which is not taxed. This is called a **tax allowance**.
- Tax is only paid on income earned in excess of the tax allowance. This is called **taxable income**.

  Eg 2 Tom earns £6080 per year. His tax allowance is £4895 per year and he pays tax at 10p in the £ on his taxable income. Find how much income tax Tom pays per year.

  Taxable income = $£6080 - £4895$ = £1185
  Income tax payable = $£1185 \times 0.10$ = £118.50

  > First find the taxable income, then multiply taxable income by rate in £.

- Gas, electricity and telephone bills are paid **quarterly**.
  Some bills consist of a standing charge plus a charge for the amount used.
- Money invested in a savings account at a bank or building society earns **interest**.
  Simple Interest is when the interest is paid out each year and not added to your account.
  Simple Interest = Amount invested × Time in years × Rate of interest per year.

  Eg 3 Find the Simple Interest paid on £600 invested at 5% for 6 months.

  Simple Interest = $\frac{600}{1} \times \frac{6}{12} \times \frac{5}{100} = £15$

  > 6 months = $\frac{6}{12}$ years.

- You should be able to work out a variety of problems involving personal finance.

Exercise 9

Do not use a calculator for questions 1 to 10.

1. Jenny worked $2\frac{1}{2}$ hours at £6.50 per hour. How much did she earn?

2. Amrit pays his council tax by 10 instalments.
   His first instalment is £193.25 and the other 9 instalments are £187 each.
   How much is his total council tax?

3. Sue insures her house for £180 000 and its contents for £14 000.
   The annual premiums for insurance are:
   House: 24p per annum for every £100 of cover.
   Contents: £1.30 per annum for every £100 of cover.
   What is the total cost of Sue's insurance?

4. Last year Harry paid the following gas bills.

£196.40	£62.87	£46.55	£183.06

   This year he will pay his gas bills by 12 equal monthly payments.
   Use last year's gas bills to calculate his monthly payments.

5 Travis is paid £8.67 per hour. Last week he worked 32 hours.
By using suitable approximations estimate how much he was paid for last week.
You must show all your working.

6 Esther has an annual income of £6789. She has a tax allowance of £4895.
(a) Calculate her taxable income.

She pays tax at the rate of 10p in the £ on her taxable income.
(b) How much income tax does she pay per year?

7 Alex has a part-time job. His basic rate of pay is £7.20 per hour.
After he has worked 8 hours he is paid at the overtime rate.
The overtime rate is one and a half times the basic rate.
(a) Calculate the amount that Alex is paid for one hour of overtime.
(b) Calculate Alex's total pay for a day when he worked $2\frac{3}{4}$ hours overtime after completing 8 hours at the basic rate. OCR

8 Felix is paid at time and a half for overtime. His overtime rate of pay is £8.40 per hour.
What is his basic rate of pay?

9 The treasurer of a badminton club wrote down details of the bills he expects to pay for the new season.

HIRE OF HALL – 48 NIGHTS - 3 HOURS EACH NIGHT AT £9.75 PER HOUR.
SHUTTLES – 60 BOXES AT £18.95 PER BOX.

The treasurer expects there will be 32 senior members and 13 junior members.
Last year senior members paid £92.50 and junior members paid £21.50.
The treasurer thinks that he can charge the same amount and have enough money to pay the bills. Show a rough calculation to check whether he is right. OCR

10 Ivor receives his gas bill. The charge for the gas he has used is £148 plus VAT at 5%.
(a) Calculate the VAT charged.
(b) Hence, find the total amount he has to pay.

11 For each call, a telephone company charges 15p connection fee, plus 2.5p per minute.
(a) What is the cost of a 12 minute call?
(b) Another call costs £1. How many minutes did it last? OCR

12 Angela is paid £6.90 per hour for a basic 35-hour week. Overtime is paid at time and a half.
One week Angela worked $37\frac{1}{2}$ hours. How much did Angela earn that week?

13 (a) When Jane babysits she charges £15 for an evening up to 11 pm then £2 for every quarter of an hour after 11 pm. One evening Jane earned £27.
What time did Jane finish babysitting?
(b) Jane saves £195 for 2 years at 3% simple interest.
How much interest does she receive? OCR

14 Leroy earns £13 880 per year.
He has a tax allowance of £4895 and pays tax at the rate of 10p in the £ on the first £2090 of his taxable income and 22p in the £ on the remainder.
How much income tax does he pay each year?

15 Paul invested £1200 for 2 years at a rate of 3.7% per year simple interest.
Calculate the amount of interest Paul received. OCR

16 An electricity bill is made up of two parts. VAT at 5% is added to the total.

A standing charge of £9.78 and a charge of 8.36p for each unit of electricity used.

Mr Hill has used 452 units of electricity. Calculate his electricity bill.

Ratio and Proportion

What you need to know

- The ratio 3 : 2 is read '3 to 2'.
- A ratio is used only to **compare** quantities.
  A ratio does not give information about the exact values of quantities being compared.
- Different forms of the **same ratio**, such as 2 : 1 and 6 : 3, are called **equivalent ratios**.
- In its **simplest form**, a ratio contains whole numbers which have no common factor other than 1.

Eg 1 Write £2.40 : 40p in its simplest form.
£2.40 : 40p = 240p : 40p
= 240 : 40
= 6 : 1

> All quantities in a ratio must be in the **same units** before the ratio can be simplified.

- You should be able to solve a variety of problems involving ratio.

Eg 2 The ratio of bats to balls in a box is 2 : 3.
There are 12 bats in the box.
How many balls are there?

$12 \div 2 = 6$
$2 \times 6 : 3 \times 6 = 12 : 18$
There are 18 balls in the box.

> For every 2 bats there are 3 balls.
> To find an equivalent ratio to 2 : 3, in which the first number is 12, multiply each number in the ratio by 6.

Eg 3 A wall costs £600 to build.
The costs of materials to labour are in the ratio 1 : 4.
What is the cost of labour?

$1 + 4 = 5$
£600 ÷ 5 = £120
Cost of labour = £120 × 4 = £480

> The numbers in the ratio add to 5.
> For every £5 of the total cost, £1 pays for materials and £4 pays for labour.
> So, **divide** by 5 and then **multiply** by 4.

- When two different quantities are always in the **same ratio** the two quantities are in **direct proportion**.

Eg 4 20 litres of petrol cost £14.
Find the cost of 25 litres of petrol.

20 litres cost £14
1 litre costs £14 ÷ 20 = £0.70
25 litres cost £0.70 × 25 = £17.50

> This is sometimes called the **unitary method**.
> **Divide** by 20 to find the cost of 1 litre.
> **Multiply** by 25 to find the cost of 25 litres.

Exercise 10

Do not use a calculator for questions 1 to 7.

1. Tim and Shula bought a car for £5000. Tim paid £3500 and Shula paid £1500.
   Write the ratio 3500 : 1500 as simply as possible. OCR

2. Rhys draws a plan of his classroom floor. The classroom measures 15 m by 20 m.
   He draws the plan to a scale of 1 cm to 5 m.
   What are the measurements of the classroom floor on the plan?

3. A toy box contains large bricks and small bricks in the ratio 1 : 4.
   The box contains 40 bricks. How many large bricks are in the box?

4 Amy and Brett buy a present for their mother. They share the cost in the ratio 3 : 1.
What percentage of the cost does Amy pay? OCR

5 To make mortar a builder mixes sand and cement in the ratio 3 : 1.
The builder uses 2.5 kg of cement. How much sand does he use?

6 Two girls, Erica and Sonia, shared £30 between them in the ratio 1 to 5.
How much did Sonia receive? OCR

7 Zoe is making pink paint.

(a) How much white paint does she need to mix with 50 ml of red paint?

(b) How much white paint does she need to make 300 ml of pink paint?

Pink Paint
to make 30 ml:
mix 10 ml of red paint
with 20 ml of white paint

OCR

8 Naheed is given £4. She spends £3.20 and saves the rest.
Express the amount she spends to the amount she saves as a ratio in its simplest form.

9 A pop concert is attended by 2100 people. The ratio of males to females is 2 : 3.
How many males attended the concert?

10 Dec shares a prize of £435 with Annabel in the ratio 3 : 2.
What is the difference in the amount of money they each receive?

11 The ratio of men to women playing golf one day is 7 : 3.

(a) What percentage of the people playing golf are men?

(b) There are 21 men playing. How many women are playing?

12 A recipe for fish pie includes these ingredients.

800 g	**white fish**
120 g	**prawns**
900 g	**potatoes**
150 ml	**soured cream**

(a) Richard has 1000 g of white fish.
He uses it all to make a larger pie with the recipe.
What weight of prawns should he use?

(b) Judy has 600 g of potatoes.
She uses them to make a smaller pie with the recipe.
What quantity of soured cream should she use? OCR

13 3 kg of pears cost £2.94. How much will 2 kg of pears cost?

14 Two students are talking about their school outing.

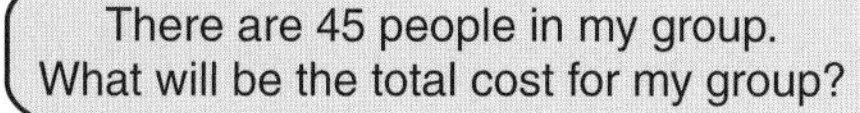

15 Every Christmas, Auntie Pat gives Emma and Rebecca money in the ratio of their ages.

(a) One year, Emma was 5 years old and Rebecca was 3 years old.
That year, Emma received £40.
How much did Rebecca receive?

(b) The next year, the girls received £90 to share between them.
How much did each girl receive? OCR

16 On a map the distance between two towns is 5 cm.
The actual distance between the towns is 1 kilometre.
What is the scale of the map in the form of 1 : n?

17 Tim, Shula and Carol share the running costs of a car in the ratio 1 : 2 : 3.
Last year, it cost £1860 to run the car. How much did Carol pay? OCR

Speed and Other Compound Measures

- **Speed** is a compound measure because it involves **two** other measures.
- **Speed** is a measurement of how fast something is travelling.
  It involves two other measures, **distance** and **time**.
  In situations where speed is not constant, **average speed** is used.

$$\text{Speed} = \frac{\text{Distance}}{\text{Time}}$$

$$\text{Average speed} = \frac{\text{Total distance travelled}}{\text{Total time taken}}$$

The formula linking speed, distance and time can be rearranged and remembered as:
$S = D \div T$
$D = S \times T$
$T = D \div S$

- You should be able to solve problems involving speed, distance and time.

Eg 1 Wyn takes 3 hours to run 24 km. Calculate his speed in kilometres per hour.

$\text{Speed} = \frac{\text{Distance}}{\text{Time}} = \frac{24}{3} = 8\text{ km/h}$

Eg 2 Norrie says, "If I drive at an average speed of 60 km/h it will take me $2\frac{1}{2}$ hours to complete my journey."
What distance is his journey?

$\text{Distance} = \text{Speed} \times \text{Time} = 60 \times 2\frac{1}{2} = 150\text{ km}$

Eg 3 Ellen cycles 5 km at an average speed of 12 km/h.
How many minutes does she take?

$\text{Time} = \frac{\text{Distance}}{\text{Speed}} = \frac{5}{12}\text{ hours} = \frac{5}{12} \times 60 = 25\text{ minutes}$

To change hours to minutes: **multiply by 60**

- **Density** is a compound measure which involves the measures **mass** and **volume**.

Eg 4 A block of metal has mass 500 g and volume 400 cm^3.

$\text{Density} = \frac{\text{Mass}}{\text{Volume}} = \frac{500}{400} = 1.25\text{ g/cm}^3$

$$\text{Density} = \frac{\text{Mass}}{\text{Volume}}$$

- **Population density** is a measure of how populated an area is.

Eg 5 The population of Cumbria is 489 700.
The area of Cumbria is 6824 km^2.

$\text{Population density} = \frac{\text{Population}}{\text{Area}} = \frac{489\,700}{6824} = 71.8\text{ people/km}^2.$

$$\text{Population density} = \frac{\text{Population}}{\text{Area}}$$

Exercise 11

Do not use a calculator for questions 1 to 5.

1. Norma travels 128 km in 2 hours.
   Calculate her average speed in kilometres per hour.

2. Sean cycled 24 km at an average speed of 16 km/h.
   How long did he take to complete the journey?

3. Ahmed takes $2\frac{1}{2}$ hours to drive from New Milton to London. He averages 66 km/h.
   What distance does he drive?

4. Nigel runs 4 km at an average speed of 6 km/h.
   How many minutes does he take?

5. Gail leaves home at 0950 to walk to the park.
She walks at an average speed of 5 km/h and reaches the park at 1020.
How far is the park from her home?

6. Paul takes 15 minutes to run to school. His average running speed is 8 km/h.
How far did he have to run?

7. Kay walks 2.5 km in 50 minutes.
Calculate her average walking speed in kilometres per hour.

8. The diagram shows the distances, in miles, between some junctions on a motorway.

West ← (25) —12— (26) —8— (27) → East

A coach is travelling west. At 1040 it passes junction 27 and at 1052 it passes junction 26.
(a) Calculate the average speed of the coach in miles per hour.

Between junctions 26 and 25 the coach travels at an average speed of 30 miles per hour.
(b) Calculate the time when the coach passes junction 25.

9. (a) Keith drives to Birmingham on a motorway.
He travels 150 miles in 2 hours 30 minutes.
Work out his average speed.
(b) He drives to Cambridge at an average speed of 57 mph.
The journey takes 3 hours 20 minutes.
How many miles is the journey? OCR

10. A train travels from Bournemouth to Manchester at an average speed of 47 miles per hour.
The train travels a distance of 268 miles.
How long does the journey take in hours and minutes?

11. Anna drove 35 miles from Southampton to Basingstoke.
She drove at an average speed of 20 mph for the first 5 miles and then at an average speed of 60 mph for the remaining 30 miles.
Calculate her average speed for the whole journey. OCR

12. A train travels at an average speed of 80 miles per hour.
At 0940 the train is 65 miles from Glasgow. The train is due to arrive in Glasgow at 1030.
Will it arrive on time? Show your working.

13. On Monday it took Helen 40 minutes to drive to work.
On Tuesday it took Helen 25 minutes to drive to work.
Her average speed on Monday was 18 miles per hour.
What was her average speed on Tuesday?

14. On a journey, Carol drove the first 90 miles. Her average speed was 60 mph.
(a) For how long did Carol drive?
(b) Tim drove the remaining 85 miles in 2 hours.
Calculate the average speed for their whole journey. OCR

15. A jet-ski travels 0.9 kilometres in 1.5 minutes.
Calculate the average speed of the jet-ski in metres per second.

16. A copper statue has a mass of 1080 g and a volume of 120 cm^3.
Work out the density of copper.

17. A silver medal has a mass of 200 g. The density of silver is 10.5 g/cm^3.
What is the volume of the medal?

18. The population of Jamaica is 2.8 million people. The area of Jamaica is 10 800 km^2.
What is the population density of Jamaica?

Section Review

Number
Non-calculator Paper

Do not use a calculator for this exercise.

1 (a) (i) Write these numbers in order of size, smallest first: 16 10 6 100 61
(ii) What is the total when the numbers are added together?
(b) Work out. (i) $100 - 37$ (ii) 100×20 (iii) $100 \div 4$

2 (a) Write the number 5031 in words.
(b) Write the number two thousand six hundred and four in figures.
(c) Write 5828 (i) correct to the nearest ten, (ii) correct to the nearest thousand.
(d) Work out the following. (i) $126 + 415$ (ii) $327 - 118$ OCR

3 (a) Copy and complete this bill.

3 metres of wood at 80p per metre.	______
200 grams of nails at 40p per 100 grams.	______
3 packets of screws at ___ per packet.	£0.75
TOTAL	£______

(b) Tessa buys a magazine costing £2.30 and a mathematics set costing £1.25.
How much change should she get from £10? OCR

4

6	8	9	11	14	15	18	27

From the list of numbers above, write down:
(a) two numbers which add up to 20,
(b) two odd numbers,
(c) a multiple of 5,
(d) the cube of 3,
(e) a prime number. OCR

5 An overnight train leaves Dundee at 2348 and arrives in London at 0735 the next day.
How long does the journey take? Give your answer in hours and minutes.

6 (a) Work out. (i) $105 - 30$ (ii) 19×7 (iii) $2002 \div 7$
(b) Work out the square of 11.
(c) Work out. (i) $8 - 3 \times 2$ (ii) $(8 - 3) \times 2$

7 (a) Write these numbers in order, starting with the smallest.

0.54	0.035	0.5	0.462	0.5089

(b) Write 0.03 as a fraction.
(c) Write 0.52 as a percentage.
(d) Write $\frac{3}{5}$ as a decimal.
(e) Which of these two is the larger? $\frac{7}{10}$ or 63%. Show how you decided. OCR

8 (a) 4 litres of milk costs £2.96. How much is 1 litre of milk?
(b) Apples cost 98 pence per kilogram. What is the cost of 5 kilograms of apples?

9 To buy a car, Ricky has to pay 24 monthly payments of £198.
How much does he have to pay altogether to buy the car?

10 Here are four number cards. **4** **5** **8** **9**
Write down the largest odd number you can make using these cards.
Each card may be used once only. OCR

11 (a) Write down all the factors of 24.
(b) What are the common factors of 24 and 36?
(c) What is the square root of 36?

12 The table shows the average minimum temperatures, in °C, in January in some cities.

Archangel	Athens	Darwin	Moscow	Ulan Bator
−20°	9°	27°	−15°	−32°

(a) What is the lowest temperature listed?
(b) What was the difference between the temperatures in Darwin and in Moscow?
(c) Hong Kong is 35°C warmer than Archangel. What is the temperature in Hong Kong?
OCR

13 A ski-run measures 7.5 cm on a map. The map is drawn to a scale of 1 cm to 200 m.
What is the actual length of the ski-run in metres?

14 How much will it cost to hire a trailer for 5 days?

15 A sports club is given £100 to spend on new footballs.
A new football costs £7.99.
What is the greatest number of footballs they can buy?

16 (a) Calculate the cost per litre of emulsion paint, correct to the nearest penny.
(b) How much more does it cost to buy 10 litres of gloss paint than 10 litres of emulsion paint?

£12.95

£14.99

17 (a) Round 3628.297 to 1 decimal place.
(b) Round 3628.297 to 1 significant figure.
OCR

18 Abdul earns £6.35 per hour. One week he worked 42 hours.
Calculate how much he earned that week.
OCR

19 (a) Work out (i) 10^5, (ii) $10^2 - 2^5$, (iii) $2^3 \times 3^2$, (iv) $30^2 \div 10^3$.
(b) Which is smaller, 5^4 or 4^5? Show **all** your working.
(c) Work out $\sqrt{25} \times \sqrt{100}$.

20 (a) Write 0.7 as a fraction.
(b) A turkey costs £3.60 per kg. What is the cost of a turkey which weighs 6.5 kg?
(c) Work out. (i) 0.2×0.4 (ii) $24 \div 0.3$

21 (a) Write these fractions in order, smallest first: $\frac{1}{2}$ $\frac{2}{3}$ $\frac{3}{5}$ $\frac{5}{8}$ $\frac{3}{4}$
(b) Write down a fraction that lies halfway between $\frac{1}{5}$ and $\frac{1}{4}$.
(c) Work out (i) $\frac{1}{4} + \frac{2}{5}$, (ii) $\frac{2}{3} - \frac{1}{2}$, (iii) $\frac{4}{5} \times \frac{2}{3}$.
(d) Work out $\frac{2}{5}$ of 12.

22 A crowd of 54 000 people watch a carnival.
(a) 15% of the crowd are men. How many men watch the carnival?
(b) Two-thirds of the crowd are children. How many children watch the carnival?

23 (a) Given that $59 \times 347 = 20\,473$, find the exact value of $\frac{20\,473}{590}$.
(b) Use approximations to estimate the value of 49×302. Show all your working.

24 Some vans and cars are parked in a car park. The ratio of vans to cars is 1 : 6.
There are 30 vans. How many cars are there?
OCR

25 (a) Zac earns £45 per week for a part-time job.
He spends 60% of his earnings on driving lessons.
How much does he spend each week on driving lessons?

(b) Anita's take-home pay is £720 per month.
She gives her mother $\frac{1}{3}$ of this and spends $\frac{1}{5}$ of the £720 on clothes.
What fraction of the £720 is left? Give your answer as a fraction in its lowest terms. OCR

26 Eric puts £350 into a savings account which pays 5% simple interest per year.
How much interest will there be after 4 years? OCR

27 (a) Diesel costs £0.95 per litre in England. Calculate the cost of 45 litres of diesel.

(b) In France, diesel is 20% cheaper than in England.
Calculate the cost of 45 litres of diesel in France.

28 A teacher has £1500 to spend on new books.
He wants 52 novels costing £9.95 each, 96 poetry books costing £6.99 each and 32 books of modern plays costing £14.75 each. He thinks he has enough money to buy all these books.
Show a rough calculation to check whether he is right. OCR

29 In a parish council election, there were three candidates, Mrs Pearson, Mr Royal and Mrs Amos. A total of 320 people voted.

(a) There were 120 votes for Mrs Pearson.
What fraction voted for Mrs Pearson? Give your answer in its lowest terms.

(b) There were 80 votes for Mr Royal. What percentage voted for Mr Royal? OCR

30 (a) Conrad cycles 24 km in $1\frac{1}{2}$ hours. What is his cycling speed in kilometres per hour?

(b) Cas cycles 24 km at 15 km/h. She sets off at 0930. At what time does she finish?

31 A box of straws contains 250 straws, to the nearest 10.
Write down the smallest and largest possible number of straws in the box.

32 (a) Given that $46.2 \times 127 = 5867.4$, write down the value of:
(i) 462×1270, (ii) $\frac{58.674}{127}$.

(b) Estimate the value of $\frac{3062}{52 \times 19}$. Show all the approximations you make. OCR

33 (a) Find. (i) the cube root of 64 (ii) the reciprocal of 0.1

(b) Work out. $1\frac{2}{3} + 2\frac{3}{4}$ OCR

34 Write as a single power of 3: (a) $3^4 \times 3^3$ (b) $3^{10} \div 3^5$ (c) $\frac{3 \times 3^3}{3^2}$

35 (a) Work out $5^2 \times 2^3$.

(b) Write 30 as a product of prime factors. OCR

36 (a) In a sale, raspberries are reduced from £5 to £4 per kg and blackberries reduced from £4.50 to £3.50 per kg.
Without doing any calculation, explain which is the greater proportional reduction.

(b) In a summer fruit pudding, raspberries and blackberries are used in the ratio 4 : 1.
A pudding contains a total of 150 g of raspberries and blackberries.
What weight of raspberries does the pudding contain?

(c) The raspberries were reduced from £5 to £4 per kg.
Calculate the percentage decrease in the cost of the raspberries.

(d) A tray of raspberries weighs 9 kg, correct to the nearest kg.
What is the least weight that the tray of raspberries could be? OCR

37 A youth club organises a skiing holiday for 45 children. The ratio of boys to girls is 5 : 4.
40% of the boys have skied before. How many boys have skied before?

38 (a) Write as a product of its prime factors: (i) 48, (ii) 108.

(b) Hence, find the least common multiple of 48 and 108.

Number
Calculator Paper

You may use a calculator for this exercise.

1 (a) Which of the numbers 8, −4, 0 or 5 is an odd number?
(b) Write the number 3568 to the nearest 10.
(c) What is the value of the 4 in the number 3.42?

2 (a) List these numbers in order, smallest first.

13	5	−7	0	−1

(b) What is the difference between the largest number and the smallest number in your list?

3 This is a timetable for travelling to Leeds Castle from London by train and coach.

Train from London	*dep.*	0908	1008	1108	1208	1308
Bearsted	*arr.*	1003	1103	1203	1303	1403
Coach from Bearsted	*dep.*	1030	1130	1230	1330	1430
Leeds Castle	*arr.*	1050	1150	1250	1350	1450

(a) Darren is travelling to Leeds Castle. He gets the train from London at 0908.
How long must he wait for the coach at Bearsted?
(b) Alice catches the 1108 train from London.
(i) At what time does she get to Leeds Castle?
(ii) How long does it take her to get from London to Leeds Castle?
(c) A travel company charges £21.80 for a day trip to Leeds Castle. Pensioners pay half price.
How much does a pensioner pay? OCR

4 Isaac buys 180 grams of sweets from the Pic 'n' Mix selection.
The price of the sweets is 65p per 100 g.
How much does he have to pay?

5 (a) A shop sells DVDs for £13.50 each.
How many DVDs can you buy for £100 and how much change should you get?
(b) Stickers cost 12p for a packet.
Aaron buys some packets of stickers and pays with a £1 coin. He gets 28p change.
How many packets of stickers did he buy? OCR

6 Here is a list of numbers: 4 10 6 24 12 22 36
(a) Which of these numbers are factors of 30?
(b) Find the biggest number that can be obtained by multiplying two numbers on the list.
OCR

7 A newspaper claimed: **ONE MOBILE PHONE SOLD EVERY FOUR SECONDS**

(a) At this rate, how many mobile phones were sold each minute?
(b) At this rate, how many mobile phones were sold in a whole year?
Give your answers to the nearest million. OCR

8 (a) Write $\frac{7}{9}$ as a decimal. Give your answer correct to two decimal places.
(b) Write 33%, 0.3, $\frac{8}{25}$ and $\frac{1}{3}$ in order of size, smallest first.

9 Ahmed went shopping and bought 2 loaves at 39p each, one chicken for £4.80,
3 kg of potatoes at 32p a kg and some tins of beans at 15p each.
He paid with a £20 note and received £13.01 change.
How many tins of beans did he buy? OCR

10 A shop makes a special offer on fertilizer.

Original price	£3.68
Sale price	£2.85

(a) By how much is the original price reduced?
(b) The fertilizer still does not sell.
The shopkeeper decides to offer a 75% reduction on the original price of £3.68.
What is the new price of the fertilizer? OCR

11 Bertie has to work out $4.2 \times 4.9 \times 31$. He uses a calculator and gets 6379.8
(a) By rounding each number to one significant figure check Bertie's answer.
Show all your working.
(b) What is the mistake in Bertie's answer?

12 At Pontmoor School there are 270 students in year 11.
Of these, $\frac{5}{9}$ are entered for GCSE Geography.
(a) How many students are entered for GCSE Geography?

Of the 270 students, 151 are girls.
(b) What percentage of the students are girls? OCR

13 Kathryn and Matt went on holiday to Spain.
(a) Before they left, they changed £200 into euros.
The rate of exchange was: £1 = 1.45 euros. How many euros did they receive?
(b) When they returned, they changed 40 euros back into pounds.
The rate of exchange was: £1 = 1.48 euros. How much did they receive?

14 Mumtaz is paid at a basic rate of £7.20 per hour for the first 37 hours she works in a week.
For any hours she works above 37, she is paid overtime.
The overtime rate is one and a half times the basic rate.
Calculate how much she earns for a week in which she works 46 hours. OCR

15 Jacob is 3.7 kg heavier than Isaac. The sum of their weights is 44.5 kg. How heavy is Jacob?

16 On a musical keyboard there are 5 black keys for every 7 white keys.
The keyboard has 28 white keys. How many black keys does it have?

17 (a) Write the number 52.03718 correct to:
(i) 3 decimal places (ii) 2 decimal places (iii) 1 decimal place
(b) Use your calculator to work these out.
(i) $\sqrt{44.89}$ (ii) 2.9^2 (iii) $11 + 4.5 \times 12$ (iv) $\frac{8}{10.3 - 7.9}$ OCR

18 Mrs Joy's electricity meter was read on 1st March and 1st June.
On 1st March the reading was 3 2 4 5 7 On 1st June the reading was 3 2 9 3 1
(a) How many units of electricity have been used?

Her electricity bill for this period includes a fixed charge of £11.58 and the cost of the units used at 9.36 pence per unit.
(b) Calculate the total cost of electricity for this period.

19 To make 16 Viennese shortcakes you need:

200 g of flour	**80 g of icing sugar**	**A little red jam**
225 g of butter	**50 g of cornflour**	

(a) Mary made 40 Viennese shortcakes.
How much icing sugar did she use?
(b) Gretchen is going to make some Viennese shortcakes. She has 375 g of flour.
How many can she make?
(c) The ingredients for the 16-cake recipe cost £1.12 in total. The cakes are to be sold at a fete.
How much should be charged for each cake to make 85% profit on the cost?
Give your answer correct to the nearest penny. OCR

20 560 people live in a village. Of these, 320 are adults and 240 are children.
(a) Calculate, in its lowest terms, the ratio of adults to children.
(b) The village stages a pantomime.
Half of the adults and three quarters of the children go to see it.
What percentage of the total population of the village see the pantomime?
(c) A pantomime ticket costs £3.50 for an adult. The price is reduced by 40% for a child.
Calculate the cost of a pantomime ticket for a child. OCR

21 The diagram shows the weights and prices of two packets of gravy granules.
This week both packets are on special offer.
The smaller packet has one third off the normal price.
The larger packet has 30% off the normal price.
Which packet is better value this week? Show your working.

22 To make squash, orange juice and water is mixed in the ratio of 1 : 6.
How much orange juice is needed to make 3.5 litres of squash?

23 A train took 3 hours 30 minutes to travel the 238 miles from London to Preston.
Calculate, in miles per hour, the average speed of the train. OCR

24 Ruby buys a new exhaust for her car. The cost is £98 plus $17\frac{1}{2}$% VAT.
How much does she have to pay altogether?

25 (a) Use your calculator to work this out. $\frac{1}{2} + \frac{10}{0.625}$
(b) Write 0.7, 11%, 0.08 and $\frac{9}{20}$ in ascending order of size.
Show working to support your answer.
(c) 1.2 kg of tomatoes and 0.3 kg of mushrooms cost a total of £1.95.
Tomatoes cost £1.10 per kilogram.
Find the cost of 1 kg of mushrooms. OCR

26 A caravan is for sale at £7200. Stuart buys the caravan on credit.
The credit terms are:

deposit 25% of sale price and 36 monthly payments of £175.

Express the extra amount paid for credit, compared with the cash price, as a percentage of the cash price.

27 Josie invests £800 in an account that pays her 3% simple interest every year.
How much interest will she have been paid in total after 6 years? OCR

28 (a) (i) Calculate $\frac{612 \times 29.6}{81.3 - 18.9}$, correct to 3 significant figures.
(ii) Use approximations to show that your answer is about right.
(b) What is the reciprocal of 0.25?

29 Rasheed, Kerry and Anthony each buy heating oil at the same price per litre.
Rasheed bought 650 litres of oil for £286.
(a) Kerry bought 800 litres of oil. How much did she pay?
(b) Anthony paid £506. How much oil did he buy? OCR

30 Calculate (a) $\frac{1}{0.72 + 0.88}$, (b) $(4.5)^2 + (3.5)^3$. OCR

31 (a) James and Anne bought a new car for £14 756. They shared the cost in the ratio 4 : 3.
How much did Anne pay?
(b) The next year the price increased from £14 756 to £15 999.
Calculate the percentage increase. OCR

32 $p = 3^2 \times 5 \times 7$ and $q = 2 \times 3 \times 5^2$. Find the least common multiple of p and q.

SECTION 12 Introduction to Algebra

What you need to know

- You should be able to write **algebraic expressions**.

  Eg 1 An expression for the cost of 6 pens at n pence each is $6n$ pence.

  Eg 2 An expression for 2 pence more than n pence is $n + 2$ pence.

- Be able to **simplify expressions** by collecting **like terms** together.

  Eg 3 (a) $2d + 3d = 5d$ (b) $3x + 2 - x + 4 = 2x + 6$ (c) $x + 2x + x^2 = 3x + x^2$

- Be able to **multiply expressions** together.

  Eg 4 (a) $2a \times a = 2a^2$ (b) $y \times y \times y = y^3$ (c) $3m \times 2n = 6mn$

- Recall and use these properties of powers:
  Powers of the same base are **added** when terms are **multiplied**.
  Powers of the same base are **subtracted** when terms are **divided**.

  $a^m \times a^n = a^{m+n}$
  $a^m \div a^n = a^{m-n}$

  Eg 5 (a) $x^3 \times x^2 = x^5$ (b) $a^5 \div a^2 = a^3$

- Be able to **multiply out brackets**.

  Eg 6 (a) $2(x + 5) = 2x + 10$ (b) $x(x - 5) = x^2 - 5x$
  (c) $(x + 2)(x + 5) = x^2 + 5x + 2x + 10 = x^2 + 7x + 10$

- Be able to **factorise expressions**.

  Eg 7 (a) $3x - 6 = 3(x - 2)$ (b) $m^2 + 5m = m(m + 5)$

Exercise 12

1. A calculator costs £9.
   Write an expression for the cost of k calculators.

2. Godfrey is 5 years older than Mary.
   Write an expression for Godfrey's age when Mary is t years old.

3. Simplify. (a) $8a + 3a - 9a$ (b) $6c + 4d - 2d - c$ OCR

4. A cup of coffee costs x pence and a cup of tea costs y pence.
   Write an expression for the cost of 3 cups of coffee and 2 cups of tea.

5. Simplify. (a) $m + 2m + 3m$ (b) $2m + 2 - m$ (c) $m \times m \times m$

6. Simplify. $4x - 2y + 3x + 3y$ OCR

7. Write an expression, in terms of x,
   for the sum of the angles in this shape.

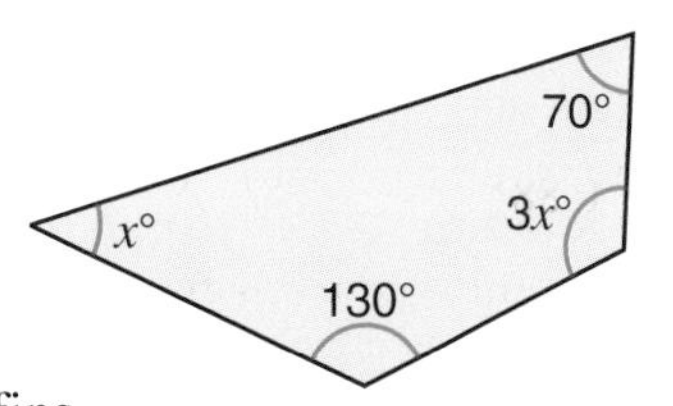

8. A muffin costs $d + 3$ pence.
   Write an expression for the cost of 5 muffins.

9. (a) Simplify. $3p + 5t + 7 - 2p + t + 9$
   (b) Multiply out the brackets. $2(3p - 5t)$ OCR

10 This shape is made from equilateral triangles.
Each equilateral triangle has an area, E, and sides of length s.

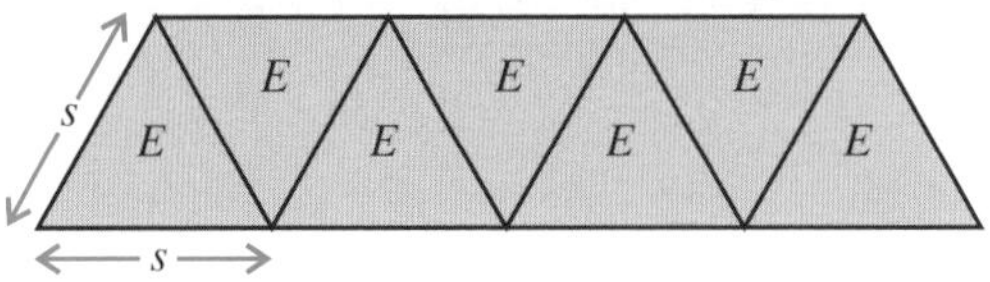

(a) Write down an expression for the area of the whole shape.
(b) Write down an expression for the perimeter of the whole shape. OCR

11 Which of these algebraic expressions are equivalent?

$a + a$	$2(a + 1)$	$2a + 1$	$2a + 2$	a^3
a^2	$a + a + 1$	$2a$	$a + a + a$	$a \times a$

12 (a) Simplify (i) $2x + 3 + x$, (ii) $2x + y - x + y$.
(b) Multiply out (i) $2(x + 3)$, (ii) $x(x - 1)$.
(c) Multiply out and simplify (i) $2(x - 1) - 3$, (ii) $7 + 3(2 + x)$.
(d) Factorise (i) $2a - 6$, (ii) $x^2 + 2x$.

13 Simplify (a) $5a \times 2a$, (b) $3g \times 2h$, (c) $6k \div 3$, (d) $3m \div m$.

14 (a) Ken works x hours a week for £y per hour.
Write an expression for the amount he earns each week.
(b) Sue works 5 hours less than Ken each week and earns £y per hour.
Write an expression for the amount Sue earns each week.

15 Bananas cost x pence each.
I buy 6 bananas and pay with a £5 note.
Write down, in terms of x, how much change I will get.
Give your answer in pence. OCR

16 (a) Simplify $2ab + 3a - 2b + b - 5a + ab$.
(b) Multiply out and simplify $3(2x + 3) + 2(5 + x)$.

17 Simplify. (a) $g + 2g + 3g$ (b) $h^3 \times h^5$ (c) $m^4 \div m$ OCR

18 (a)

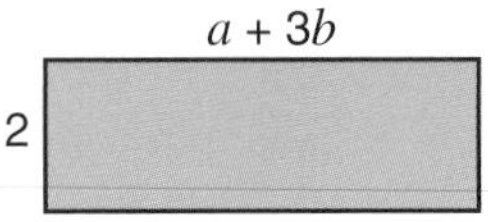

This rectangle has area $2(a + 3b)$.
Multiply out $2(a + 3b)$.

(b) This rectangle has area $3a + 12$.
The width of the rectangle is 3.
Write down an expression for the length of the rectangle.

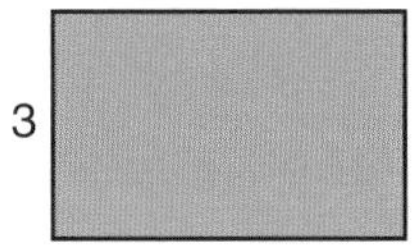

OCR

19 Expand and simplify $3(2x + 3) - 2(5 + x)$.

20 Simplify. (a) $y^3 \times y^2$ (b) $x^6 \div x^3$ (c) $\dfrac{z^4 \times z}{z^3}$

21 Multiply out the brackets and simplify your answer. $(x - 3)(x + 5)$ OCR

22 Simplify. (a) $a \times a \times a \times a$ (b) $b \times b^3$ (c) $\dfrac{c^5}{c^2}$ (d) $\dfrac{d^3 \times d^5}{d^4}$

23 (a) Factorise $4x + 6$.
(b) Expand (i) $3(2y - 3)$, (ii) $x(x^2 - 2x)$, (iii) $a(a + b)$.
(c) Simplify. $2x^2 - x(1 + x)$

24 Expand and simplify. $(m - 2)(m - 3)$

SECTION 

Solving Equations

What you need to know

- The solution of an equation is the value of the unknown letter that fits the equation.
- You should be able to solve simple equations by **inspection**.

Eg 1 (a) $a + 2 = 5$ (b) $m - 3 = 7$ (c) $2x = 10$ (d) $\frac{t}{4} = 3$

$a = 3$ $m = 10$ $x = 5$ $t = 12$

- Be able to solve simple problems by **working backwards**.

Eg 2 I think of a number, multiply it by 3 and add 4. The answer is 19.

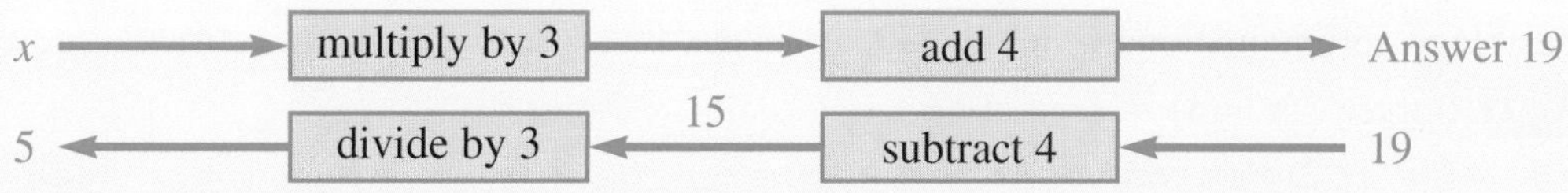

The number I thought of is 5.

- Be able to use the **balance method** to solve equations.

Eg 3 Solve these equations.

(a) $d - 13 = -5$

$d = -5 + 13$

$d = 8$

(b) $-4a = 20$

$a = \frac{20}{-4}$

$a = -5$

(c) $5 - 4n = -1$

$-4n = -6$

$n = 1.5$

Exercise 13

1 What number should be put in the box to make each of these statements correct?

(a) $\square - 6 = 9$ (b) $2 + \square = 11$ (c) $4 \times \square = 20$ (d) $\frac{\square}{5} = 3$

2 Solve. (a) $a + 2 = 7$ (b) $10 - b = 8$ (c) $3c = 12$ OCR

3 Andy thought of a number. He multiplied the number by 2. His answer was 18.
What number did he think of?

4 The diagram shows a mathematical rule.
Copy and complete the table.

Input → × 2 → + 3 → Output

Input	3		−2
Output		13	

5 (a) I think of a number, add 3, and then multiply by 2. The answer is 16. What is my number?
(b) I think of a number, double it and then subtract 3. The answer is 5. What is my number?

6 Solve. (a) $6x = 30$ (b) $15 = x - 8$ (c) $5x + 1 = 38$ OCR

7 Solve these equations.
(a) $3x - 7 = 23$ (b) $4 + 3x = 19$ (c) $5x - 9 = 11$ (d) $5 - 7x = 47$

8 Solve these equations.
(a) $3x + 5 = 2$ (b) $4x = 2$ (c) $4x + 1 = 23$ (d) $5x + 1 = -3$

Further Equations

What you need to know

- To solve an equation you need to find the numerical value of the letter, by ending up with **one letter** on one side of the equation and a **number** on the other side of the equation.
- You should be able to solve equations with unknowns on both sides of the equals sign.

Eg 1 Solve $3x + 1 = x + 7$.

$3x = x + 6$
$2x = 6$
$x = 3$

- Be able to solve equations which include brackets.

Eg 2 Solve $2(x - 3) = 4$.

$2x - 6 = 4$
$2x = 10$
$x = 5$

- You should be able to solve equations using a **trial and improvement** method.
The value of the unknown letter is improved until the required degree of accuracy is obtained.

Eg 3 Use a trial and improvement method to find a solution to the equation $x^3 + x = 40$, correct to one decimal place.

x	$x^3 + x$	Comment
3	$27 + 3 = 30$	Too small
4	$64 + 4 = 68$	Too big
3.5	$42.8... + 3.5 = 46.3...$	Too big
3.3	$35.9... + 3.3 = 39.2...$	Too small
3.35	$37.5... + 3.35 = 40.9...$	Too big

For accuracy to 1 d.p. check the second decimal place. The solution lies between 3.3 and 3.35.

$x = 3.3$, correct to 1 d.p.

- You should be able to write, or form, equations using the information given in a problem.

Exercise 14

1. Solve these equations. (a) $3 + x = 7$ (b) $3 - x = 4$ (c) $3x = 15$ (d) $\frac{x}{3} = 7$

2. Work out the missing numbers in these calculations.

   (a) 4 → +6 → ×3 → .....

   (b) ..... → +6 → ×3 → 18

   (c) −2 → +..... → ×3 → 21

   OCR

3. x, y and z represent different numbers, such that:

   $$x + x + x = 21, \quad x + y + y = 17 \quad \text{and} \quad x + y + z = 9$$

   Find the values of x, y and z.

4. Solve. (a) $2x + 5 = 37$ (b) $6x - 1 = 1 + 2x$

   OCR

5 Solve the equations (a) $3x - 7 = x + 15$, (b) $5(x - 2) = 20$.

6 Solve these equations. (a) $\frac{x}{2} = 2$ (b) $4x + 3 = 27$ (c) $8x - 3 = 6x + 15$ OCR

7 Solve these equations.
(a) $7x + 4 = 60$ (b) $3x - 7 = -4$ (c) $2(x + 3) = -2$ (d) $3x - 4 = 1 + x$

8 Solve. (a) $5x + 4 = 2x + 8$ (b) $3(x + 10) = 24$ OCR

9 Solve these equations.
(a) $2x + 5 = 2$ (b) $2(x - 1) = 3$ (c) $5 - 2x = 3x + 2$ (d) $2(3 + x) = 9$

10 Solve $4x + 20 = 9x + 5$.

11 The lengths of these rods are given, in centimetres, in terms of n.

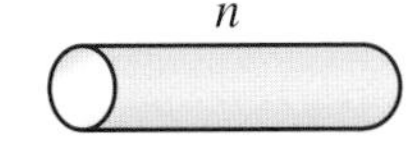

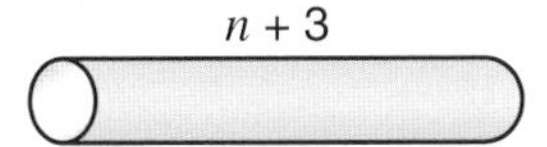

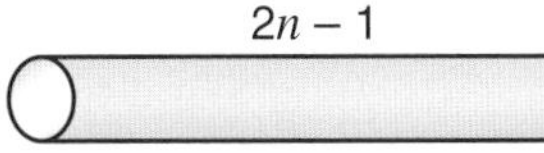

(a) Write an expression, in terms of n, for the total length of the rods.
(b) The total length of the rods is 30 cm.
By forming an equation, find the value of n.

12 Mandy buys a small box of chocolates and a large box of chocolates.
The diagram shows the number of chocolates in each box.
Altogether there are 47 chocolates.
By forming an equation, find the number of chocolates in the larger box.

13 Ice lollies cost x pence each.
Ice creams cost 90 pence each.
The total cost of three ice lollies and one ice cream is £2.85.
(a) Write down an equation in x.
(b) Solve your equation to find the cost of one ice lolly. OCR

14 Solve the equation $4(3 - x) = 20$.

15 Solve these equations. (a) $3x - 5 = 16$ (b) $7x + 1 = 2x + 4$

16 Solve the equation $5(x - 3) = 2x$.

17 A cracker costs n pence.
A party hat costs 7 pence less than a cracker.
(a) Write an expression for the cost of a party hat.
(b) The cost of 10 crackers and 5 party hats is £4.45
By forming an equation in n find the cost of a party hat.

18 (a) Expand. $x(x - 3)$
(b) Solve. $8y + 13 = 3(y + 1)$ OCR

19 Solve. (a) $3 - 4q = 11$ (b) $4(2t - 3) + 4t = 6$

20 (a) Show that one solution of the equation $x^3 - 5x - 8 = 0$ lies between 2 and 3.
(b) Use trial and improvement to find this solution, correct to one decimal place.
You must show all your trials and their outcomes. OCR

21 Use a trial and improvement method to find the value of x, correct to **one** decimal place, when $x^3 - 3x = 38$.
Show clearly your trials and their outcomes. OCR

SECTION 15

Formulae

What you need to know

- An **expression** is just an answer using letters and numbers.
  A **formula** is an algebraic rule. It always has an equals sign.
- You should be able to **write simple formulae**.

Eg 1 A packet of crisps weighs 25 grams.
Write a formula for the total weight, W grams, of n packets of crisps.
$W = 25n$

Eg 2 Start with t, add 5 and then multiply by 3.
The result is p.
Write a formula for p in terms of t.
$p = 3(t + 5)$

- Be able to **substitute** values into expressions and formulae.

Eg 3 (a) Find the value of $4x - y$ when $x = 5$ and $y = 7$.
$4x - y = 4 \times 5 - 7$
$= 20 - 7$
$= 13$

(b) $A = pq - r$
Find the value of A when $p = 2$, $q = -2$ and $r = 3$.
$A = pq - r$
$A = 2 \times (-2) - 3$
$A = -4 - 3$
$A = -7$

(c) $M = 2n^2$
Find the value of M when $n = 3$.
$M = 2n^2$
$M = 2 \times 3^2$
$M = 2 \times 9$
$M = 18$

- Be able to **rearrange** a simple formula to make another letter (variable) the subject.

Eg 4 $y = 2x + 5$. Make x the subject of the formula.
$y = 2x + 5$
Take 5 from both sides. $y - 5 = 2x$
Divide both sides by 2. $\frac{y-5}{2} = x$. So, $x = \frac{y-5}{2}$

Exercise 15

Do not use a calculator for questions 1 to 14.

1. What is the value of $a - 3b$ when $a = 10$ and $b = 2$?

2. Nikki wants to hire two machines.
These formulas are used to work out the hiring costs in pounds.

SANDER	POLISHER
multiply the number of hours by 3 then add 5	multiply the number of hours by 4 then add 6

Nikki hires the sander for 6 hours and the polisher for 2 hours.
How much does this cost altogether? OCR

3. Given that $m = -3$ and $n = 5$, find the value of
(a) $m + n$, (b) $m - n$, (c) $n - m$, (d) mn.

4. $H = ab - c$. Find the value of H when $a = 2$, $b = -5$ and $c = 3$.

5 These are formulae for a shape. **perimeter = $5a + 2b$ area = $2ab$**

Given that $a = 3$ and $b = 6$, work out (a) the perimeter, (b) the area. OCR

6 This word formula tells you how to find the cost, in pounds, of hiring a minibus.

Cost = number of miles × 2, then add 10.

(a) Steve hires a minibus to travel 30 miles to get to Luton Airport. How much will it cost?
(b) Sarah hires a minibus to travel forty-seven miles. How much will it cost?
(c) For a different journey, Dave paid £42 to hire a minibus.
How many miles was his journey? OCR

7 (a) Find the value of $3n - 1$ when the value of n is: (i) 57, (ii) -5,
(iii) Is the value of $3n - 1$ always an even number? Explain your answer.
(b) $P = 3a + 2b$. Find the value of b when $P = 27$ and $a = 5$. OCR

8 $L = 5(p + q)$. Find the value of L when $p = 2$ and $q = -4$.

9 $A = b - cd$. Find the value of A when $b = -3$, $c = 2$ and $d = 4$.

10 For the formula $F = a^2 - 3b$, find the value of F when
(a) $a = 5$, $b = 2$, (b) $a = -4$, $b = -2$. OCR

11 Work out the value of $2x^2$ when (a) $x = 3$, (b) $x = -4$. OCR

12 What is the value of $3x^3$ when $x = 2$?

13 $T = ab^2$. Find the value of T when $a = 4$ and $b = -5$.

14 Below is an extract from Reuben's homework.

Find the value of $2x^2 + 5$ when $x = -3$.
$2 \times -3^2 + 5$
Answer = 41 ✗

Explain what Reuben has done wrong. OCR

15 Eve buys n pints of milk at 35 pence a pint.
She pays for them with a £2 coin. She is given C pence change.
Write down a formula for C in terms of n.

16 This rule is used to change miles into kilometres.

Multiply the number of miles by 8 and then divide by 5

(a) Use the rule to change 25 miles into kilometres.
(b) Using K for the number of kilometres and M for the number of miles write a formula for K in terms of M.
(c) Use your formula to find the value of M when $K = 60$.

17 (a) $A = x^2 + 3x$. Find the value of A when $x = -5$.
(b) Rearrange this formula to make n the subject. $C = 10n - 5$ OCR

18 Rearrange the formula $n = 3 + mp$ to make m the subject.

19 $m = 3(n - 17)$. Find the value of n when $m = -9$.

20 Rearrange the formula $C = \frac{3r}{4}$ to make r the subject. OCR

21 You are given the formula $v = u + at$.
(a) Find v when $u = 17$, $a = -8$ and $t = 3$.
(b) Rearrange the formula to give a in terms of v, u and t.

SECTION 16

Sequences

What you need to know

- A **sequence** is a list of numbers made according to some rule.
  The numbers in a sequence are called **terms**.
- You should be able to draw and continue number sequences represented by patterns of shapes.

  Eg 1 This pattern represents the sequence: 3, 5, 7, …

- Be able to continue a sequence by following a given rule.

  Eg 2 The sequence 2, 7, 22, … is made using the rule:

  > Multiply the last number by 3, then add 1.

  The next term in the sequence $= (22 \times 3) + 1 = 66 + 1 = 67$

- Be able to find a rule, and then use it, to continue a sequence.

  > **To continue a sequence:**
  > 1. Work out the rule to get from one term to the next.
  > 2. Apply the same rule to find further terms in the sequence.

  Eg 3 Describe the rule used to make the following sequences.
  Then use the rule to find the next term of each sequence.

  (a) 5, 8, 11, 14, … Rule: add 3 to last term. Next term: 17.
  (b) 2, 4, 8, 16, … Rule: multiply last term by 2. Next term: 32.
  (c) 1, 1, 2, 3, 5, 8, … Rule: add the last two terms. Next term: 13.

- Special sequences **Square numbers:** 1, 4, 9, 16, 25, …
  Triangular numbers: 1, 3, 6, 10, 15, …
- A number sequence which increases (or decreases) by the same amount from one term to the next is called a **linear sequence**.
  The sequence: 2, 8, 14, 20, 26, … has a **common difference** of 6.
- You should be able to find an expression for the nth term of a linear sequence.

  Eg 4 Find the nth term of the sequence: 3, 5, 7, 9, …
  The sequence is linear, common difference = 2.
  To find the nth term add one to the multiples of 2.
  So, the nth term is $2n + 1$.

Exercise 16

1. Write down the next two terms in each of these linear sequences.
   (a) 1, 5, 9, 13, 17, … (b) 50, 46, 42, 38, 34, …
2. What is the next number in each of these sequences?
   (a) 1, 2, 5, 10, … (b) 1, 3, 9, 27, … (c) 1, $\frac{1}{2}$, $\frac{1}{4}$, $\frac{1}{8}$, …
3. The first six terms of a sequence are shown. 1, 4, 5, 9, 14, 23, …
   Write down the next two terms.

4 Here are the first three patterns in a sequence.

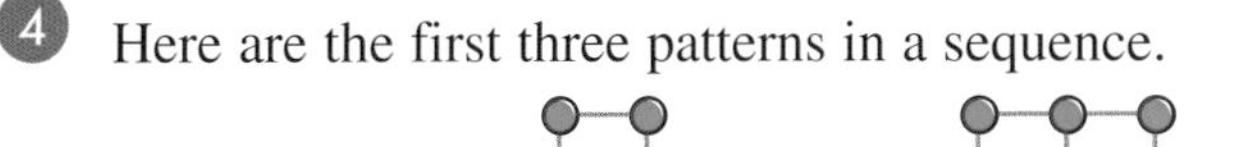

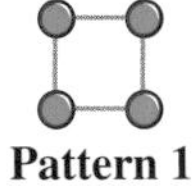

Pattern 1

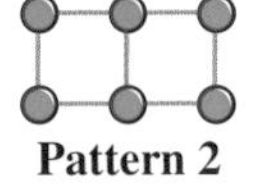

Pattern 2

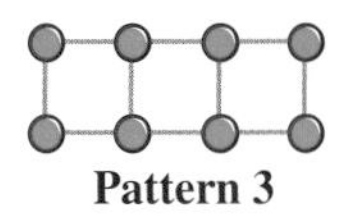

Pattern 3

(a) Draw the next pattern in the sequence.

(b) Copy and complete this table.

Pattern	1	2	3	4	5	6	7
Number of DOTS	4	6	8				
Number of LINES	4	7	10				

(c) Describe the sequence of the numbers
(i) in the middle line of your table, (ii) in the bottom line of your table.

(d) **Without drawing more diagrams**, find
(i) the number of lines in Pattern 12, (ii) which pattern has 30 dots. OCR

5 Look at this sequence of numbers. 2, 5, 8, 11, …

(a) What is the next number in the sequence?

(b) Is 30 a number in this sequence? Give a reason for your answer.

6 A sequence begins: 5, 15, 45, 135, …

(a) Write down the rule, in words, used to get from one term to the next in the sequence.

(b) Use your rule to find the next term in the sequence.

7 (a) Here is the rule for a sequence. Multiply the previous term by 3 and subtract 2.

The first term of this sequence is 4.
Write down the next two terms of this sequence.

(b) Here are the first four terms of another sequence. 128, 64, 32, 16
(i) Find the seventh term.
(ii) Explain how you worked out your answer.

(c) Here is the rule for another sequence. Subtract 4 from the previous term.

The fourth term of this sequence is 34. Find the first term. OCR

8 The first three patterns in a sequence are shown.

Pattern 1

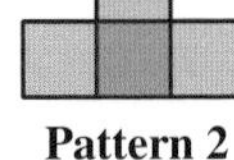

Pattern 2

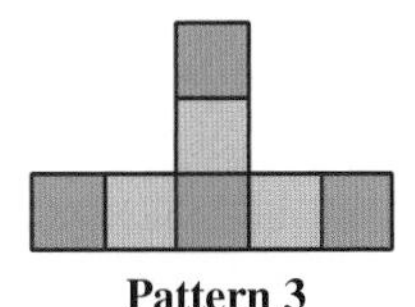

Pattern 3

(a) How many squares are in pattern 20? Explain how you found your answer.

(b) Write an expression for the number of squares in the nth pattern.

9 (a) Write down the next two terms of this sequence. 18, 17, 15, 12, 8

(b) The nth term of another sequence is: $15 - 3n$.
Find the 5th term of this sequence.

(c) These are the first five terms of another sequence. 5, 9, 13, 17, 21
Find the nth term of this sequence. OCR

10 Find the nth term of the following sequences.

(a) 5, 7, 9, 11, … (b) 1, 6, 11, 16, …

11 (a) Here are the first four terms of a sequence. 200, 136, 104, 88
Explain how to work out the next two terms.

(b) Here are the first four terms of another sequence. -1, 1, 3, 5
Find the nth term of this sequence.

(c) The nth term, T, of another sequence is given by the formula $T = n^2 - 4$.
Write down the first three terms of this sequence. OCR

SECTION 17 Coordinates and Graphs

What you need to know

- **Coordinates** (involving positive and negative numbers) are used to describe the position of a point on a graph.

  Eg 1 The coordinates of A are $(4, 1)$.
  The coordinates of B are $(-3, 2)$.

- The x axis is the line $y = 0$. The y axis is the line $x = 0$.
- The x axis crosses the y axis at the **origin**.
- The graph of a linear function is a straight line.
- You should be able to draw the graph of a straight line.

  Eg 2 Draw the graphs of the following lines.

  (a) $y = 2$

  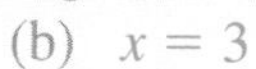

  (b) $x = 3$

  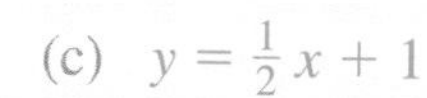

  (c) $y = \frac{1}{2}x + 1$

  The graph is a **horizontal** line. All points on the line have y coordinate 2.

  The graph is a **vertical** line. All points on the line have x coordinate 3.

  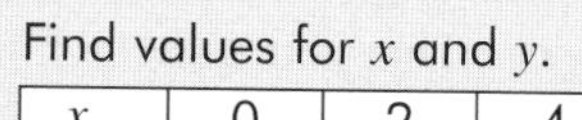

  Find values for x and y.

  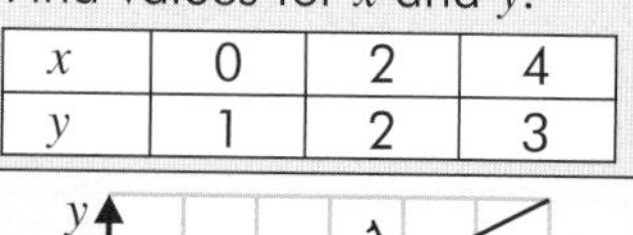

x	0	2	4
y	1	2	3

  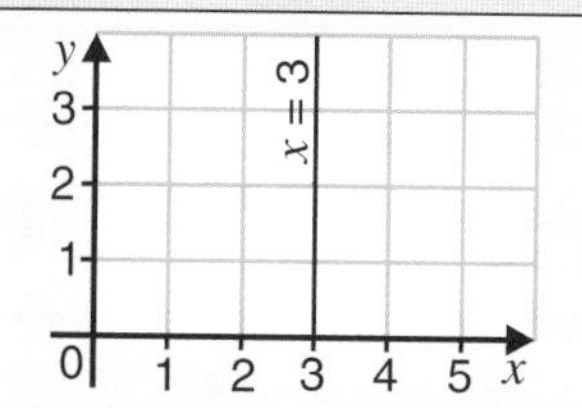

  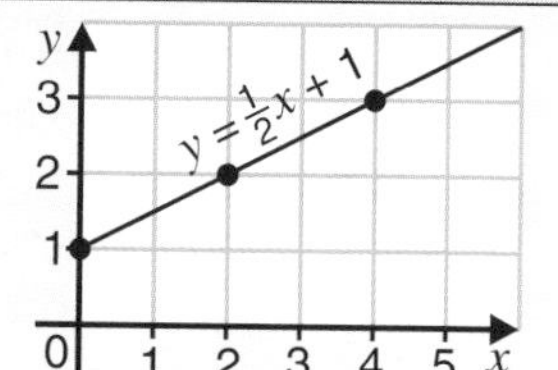

- Be able to draw the graph of a straight line by finding the points where the line crosses the x axis and the y axis.

  Eg 3 Draw the graph of the line $x + 2y = 4$.

  At the point where a graph crosses:
  the x axis, $y = 0$,
  the y axis, $x = 0$.

  When $y = 0$, $x + 0 = 4$, $x = 4$. Plot $(4, 0)$.
  When $x = 0$, $0 + 2y = 4$, $y = 2$. Plot $(0, 2)$.
  A straight line drawn through the points $(0, 2)$ and $(4, 0)$ is the graph of $x + 2y = 4$.

  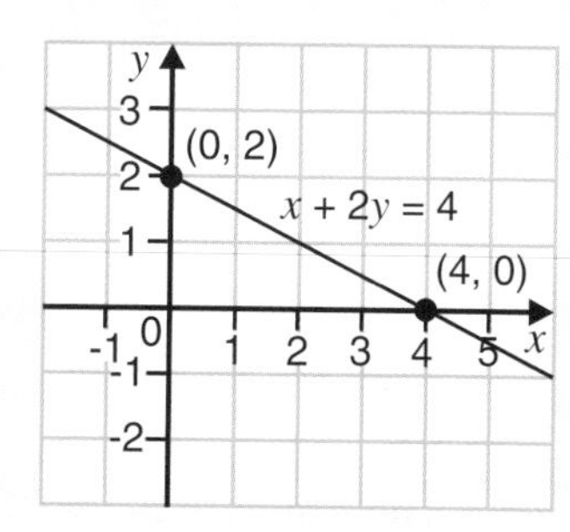

- The equation of the graph of a straight line is of the form $y = mx + c$, where m is the gradient and c is the y-intercept.

  The **gradient** of a line can be found by drawing a right-angled triangle.

  $$\text{Gradient} = \frac{\text{distance up}}{\text{distance along}}$$

  Gradients can be positive, zero or negative.

  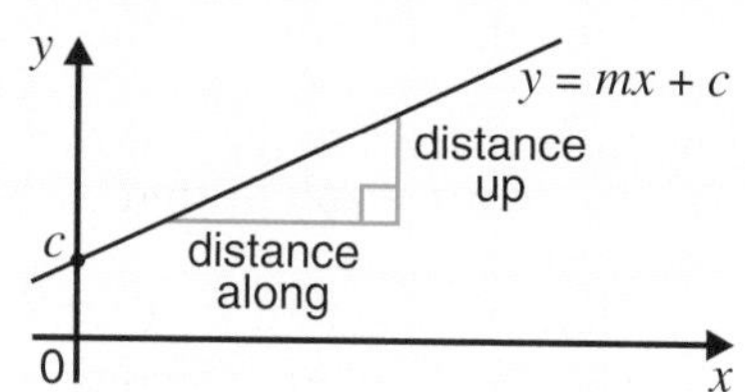

You should be able to:

- interpret the graph of a linear function,
- use the graphs of linear functions to solve equations.

Exercise 17

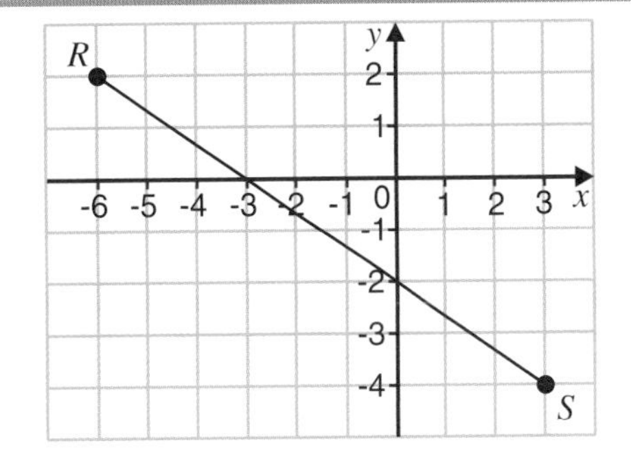

1 The diagram shows the line segment RS.

(a) Write down the coordinates of points R and S.

(b) The straight line joining R and S crosses the x axis at T and the y axis at U. Write down the coordinates of T and U.

2 Draw and label x and y axes from -5 to 4.

(a) On your diagram plot $A\,(4, 3)$ and $B\,(-5, -3)$.

(b) $C\,(p, -1)$ is on the line segment AB. What is the value of p?

3 (a) On the same diagram draw the lines $y = 2$ and $x = 5$.

(b) Write down the coordinates of the point where the lines cross.

4 (a) Complete the table below for $y = 2x - 4$.

x	0	1	2	3	4	5
y	-4		0	2		6

(b) Draw the graph of $y = 2x - 4$. OCR

5 (a) On the same axes, draw the graphs of $y = -2$, $y = x$ and $x + y = 5$.

(b) Which of these lines has a negative gradient?

6 Draw the graph of $y = 3x - 2$. Use values of x from 0 to 5. OCR

7 (a) Draw the graph of $y = 1 - 2x$ for values of x from -3 to 3.

(b) Use your graph to find the value of y when $x = -1.5$.

8 The diagram shows a sketch of the line $2y = 6 - x$.

(a) Find the coordinates of the points P and Q.

(b) The line $2y = 6 - x$ goes through $R\,(-5, m)$. What is the value of m?

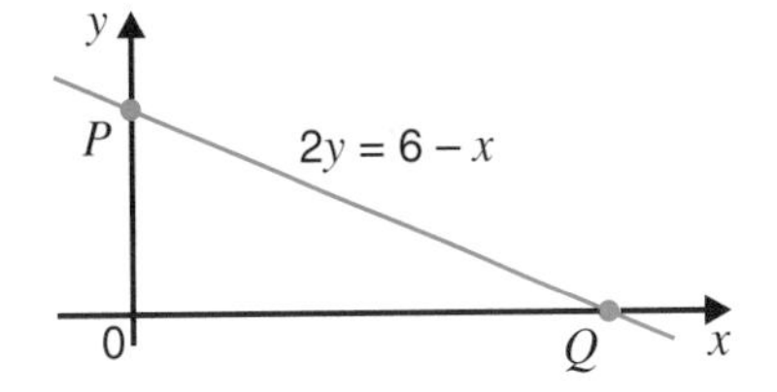

9 Points A, B and C are shown on the grid.

(a) Write down the equation of the line AB.

(b) Use the grid to work out the equation of the line CB.

10 (a) Copy and complete the table of values for $2y = 3x - 6$.

x	-2	0	4
y		-3	

(b) Draw the graph of $2y = 3x - 6$ for values of x from -2 to 4.

(c) Use your graph to find the value of x when $y = 1.5$.

11 (a) Draw the graph of $5y - 2x = 10$ for values of x from -5 to 5.

(b) Use your graph to find the value of y when $x = -2$.

12 (a) On the same diagram, draw and label the lines $y = x - 1$ and $x + y = 4$ for values of x from 0 to 4.

(b) Write down the coordinates of the point where the lines cross.

SECTION 18 Using Graphs

What you need to know

- A graph used to change from one quantity into an equivalent quantity is called a **conversion graph**.

Eg 1 Use 15 kilograms = 33 pounds (lb) to draw a conversion graph for kilograms and pounds.
Use your graph to find (a) 5 kilograms in pounds, (b) 20 pounds in kilograms.

The straight line drawn through the points (0, 0) and (33, 15) is the conversion graph for kilograms and pounds.

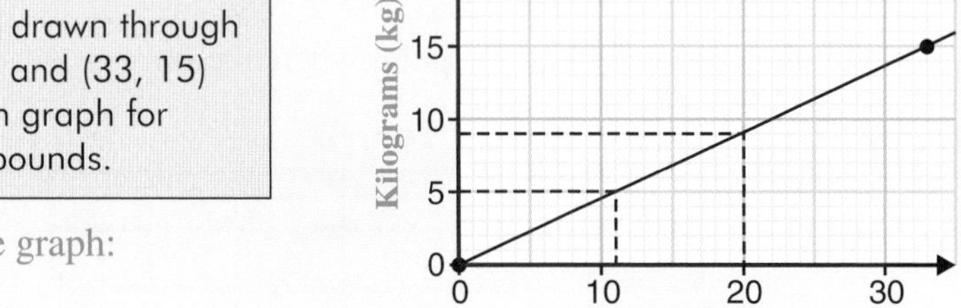

Reading from the graph:
(a) 5 kg = 11 lb
(b) 20 lb = 9 kg

- **Distance-time graphs** are used to illustrate journeys.

On a distance-time graph:
Speed can be calculated from the gradient of a line.
The faster the speed the steeper the gradient.
Zero gradient (horizontal line) means zero speed.

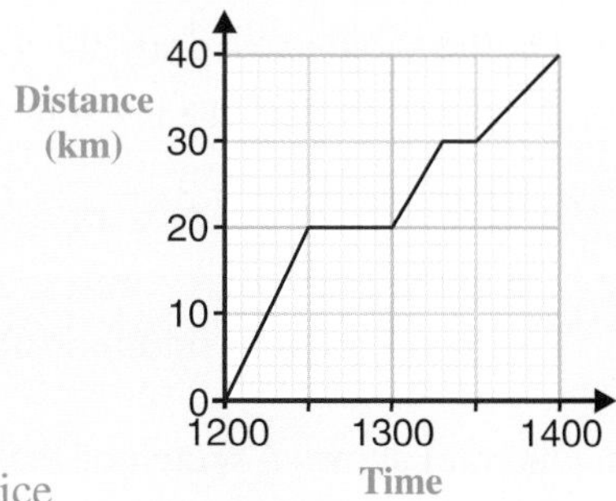

Eg 2 The graph shows a car journey.
(a) How many times does the car stop?
(b) (i) Between what times does the car travel fastest? Explain your answer.
(ii) What is the speed of the car during this part of the journey?

(a) Twice
(b) (i) 1200 to 1230. Steepest gradient.
(ii) Speed $= \frac{\text{Distance}}{\text{Time}} = \frac{20\,\text{km}}{\frac{1}{2}\text{ hour}} = 40\,\text{km/h}$

- You should be able to draw and interpret graphs arising from real-life situations.

Exercise 18

1. This conversion graph can be used to convert between miles and kilometres.

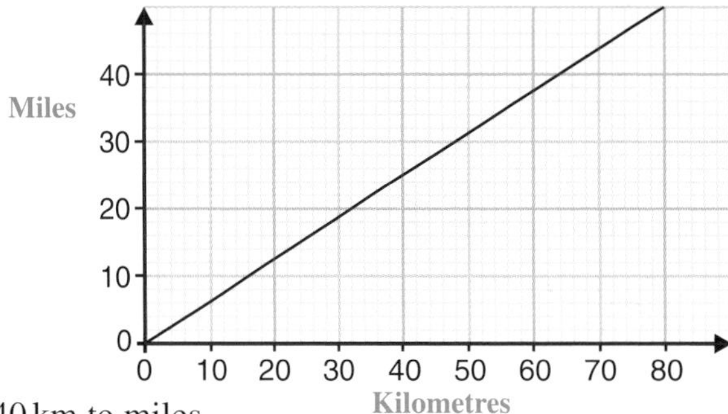

(a) Convert 40 km to miles.
(b) Convert 10 miles to kilometres.
(c) Explain how the graph could be used to convert 500 miles to kilometres.

2 The graph shows the temperature of the water in a tank as it is being heated.

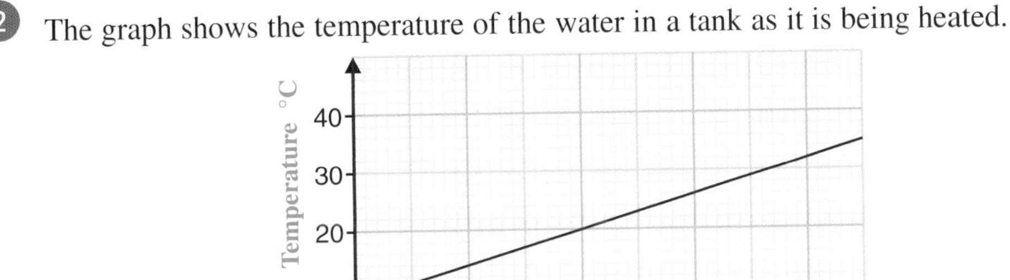

(a) What was the temperature of the water before it was heated?
(b) How long did it take for the water to reach 26°C?
(c) Estimate the number of minutes it will take for the temperature of the water to rise from 32°C to 50°C.

3 (a) Given that 7.4 square metres = 80 square feet, draw a conversion graph for square metres to square feet.
(b) Use your graph to change
(i) 5 square metres into square feet,
(ii) 32 square feet into square metres.

4 'Anywhere' Car Rental has two rates, A and B, for the daily hire of their cars.
The graph for Rate A is shown

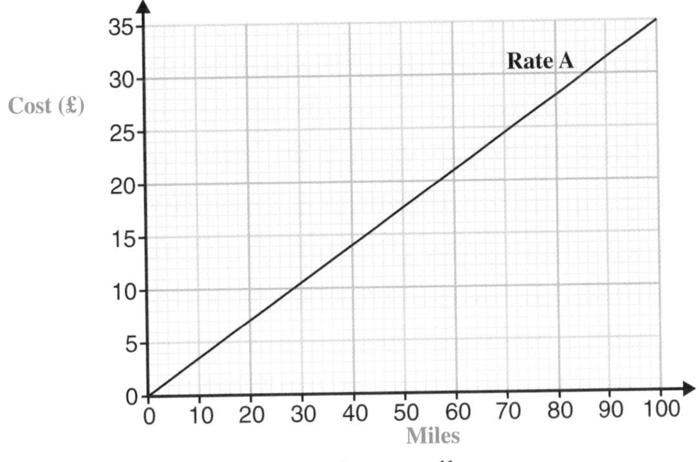

(a) Rate B is £20 fixed charge plus 10p per mile.
Copy the graph for Rate A and draw the graph for Rate B on the same grid.
(b) For what mileage is the cost of hiring a car the same for Rate A and Rate B? OCR

5 Hassan is driving along a motorway. The distance-time graph shows part of his journey.

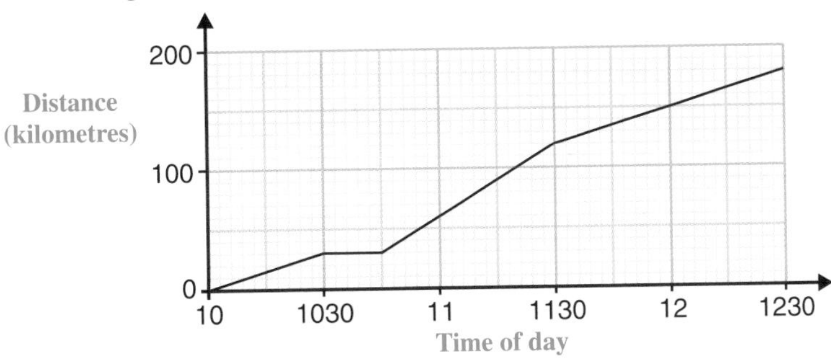

(a) Between which times is Hassan driving fastest?
(b) Calculate Hassan's speed between 10 am and 10.30 am. OCR

6 A salesman is paid a basic amount each month plus commission on sales.
The graph shows how the monthly pay of the salesman depends on his sales.

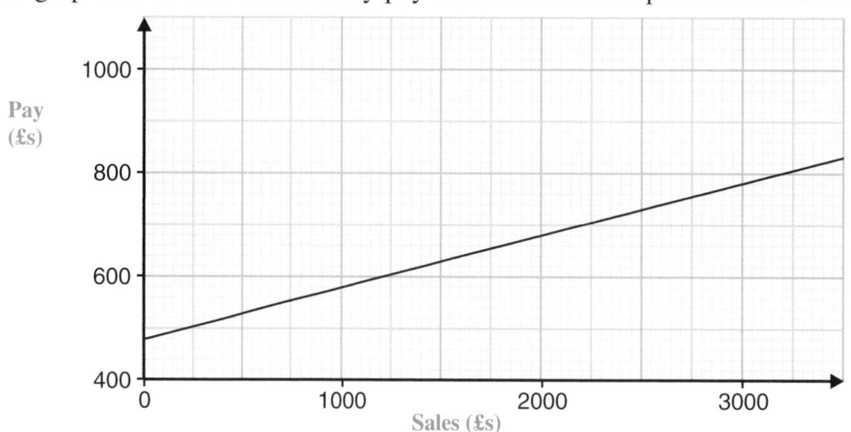

(a) How much is the salesman's monthly basic pay?
(b) How much commission is the salesman paid on £1000 of sales?
(c) Calculate the monthly pay of the salesman when his sales are £5000.

7 The graph shows Ahmed's journey. He went to the post box then on to the minimarket.

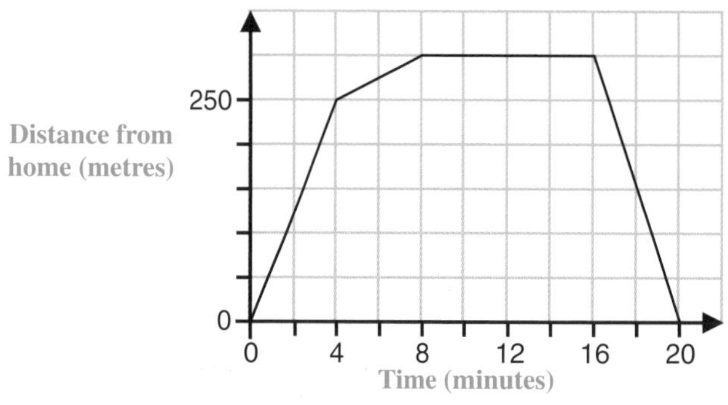

(a) The post box is 250 m from Ahmed's home.
How long did it take Ahmed to walk to the post box?
(b) How far is the minimarket from Ahmed's home?
(c) The last section of the graph slopes down.
What does this show?

OCR

8 Steve travelled from home to school by walking to a bus stop and then catching a school bus.
(a) Use the information below to construct a travel graph showing Steve's journey.
Steve left home at 0800.
He walked at 6 km/h for 10 minutes.
He then waited for 5 minutes before catching the bus.
The bus took him a further 8 km to school at a steady speed of 32 km/h.
(b) How far is Steve from home at 0820?
(c) (i) How long would it take Steve to cycle from home to school at an average speed of 15 km/h? Give your answer in minutes.
(ii) Steve cycles at 15 km/h and wants to arrive at the same time as the bus in part (a).
At what time must he leave home?

OCR

9 Water is poured at a constant rate into each of the two containers shown.
Sketch graphs to show the height of water in the containers as they are being filled.

(a)

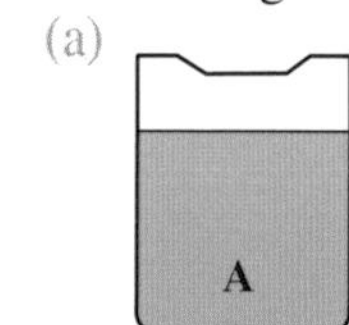

(b)

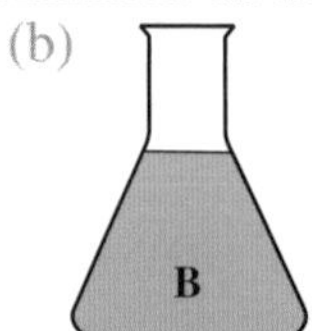

SECTION 19

Inequalities

What you need to know

- **Inequalities** can be described using words or numbers and symbols.

Sign	Meaning
$<$	is less than
$\leqslant$	is less than or equal to

Sign	Meaning
$>$	is greater than
$\geqslant$	is greater than or equal to

- Inequalities can be shown on a **number line**.

Eg 1 This diagram shows the inequality: $-2 < x \leqslant 3$

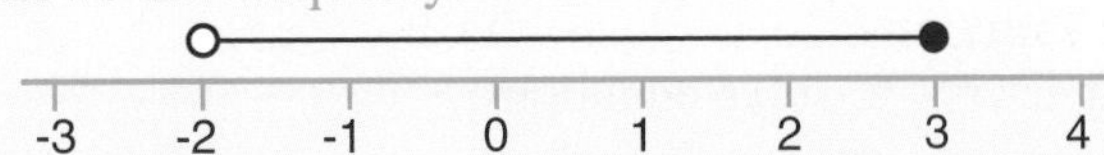

The circle is: **filled** if the inequality is **included** (i.e. $\leqslant$ or $\geqslant$),
not filled if the inequality is **not included** (i.e. $<$ or $>$).

- **Solving inequalities** means finding the values of x which make the inequality true.

Eg 2 Solve these inequalities.

(a) $3x < 6$
$x < 2$

(b) $x + 3 \geqslant 5$
$x \geqslant 2$

(c) $7a \geqslant a + 9$
$6a \geqslant 9$
$a \geqslant 1.5$

Eg 3 Find the integer values of n for which $-1 \leqslant 2n + 3 < 7$.

$-1 \leqslant 2n + 3 < 7$
$-4 \leqslant 2n < 4$
$-2 \leqslant n < 2$

Integer values which satisfy the inequality $-1 \leqslant 2n + 3 < 7$ are: -2, -1, 0, 1

Exercise 19

1 Solve these inequalities.
(a) $5x > 15$ (b) $x + 3 \geqslant 1$ (c) $x - 5 \leqslant 1$ (d) $3 + 2x > 7$

2 Draw number lines to show each of these inequalities.
(a) $x \geqslant -2$ (b) $\frac{x}{3} < -1$ (c) $-1 < x \leqslant 3$ (d) $x \leqslant -1$ **and** $x > 3$

3 (a) Solve. $2x - 1 \geqslant 5$
(b) Represent your solution to part (a) on a number line. OCR

4 Solve these inequalities.
(a) $2x \leqslant 6 - x$ (b) $3x > x + 7$ (c) $5x < 2x - 4$

5 List the values of n, where n is an integer such that:
(a) $-2 \leqslant 2n < 6$ (b) $-3 < n - 3 \leqslant -1$ (c) $-5 \leqslant 2n - 3 < 1$

6 Find all the integer values of n which satisfy $-15 < 5n \leqslant 20$. OCR

7 (a) Solve the inequality $3x - 5 \leqslant 1$.
(b) Write down the inequality shown by the following diagram.

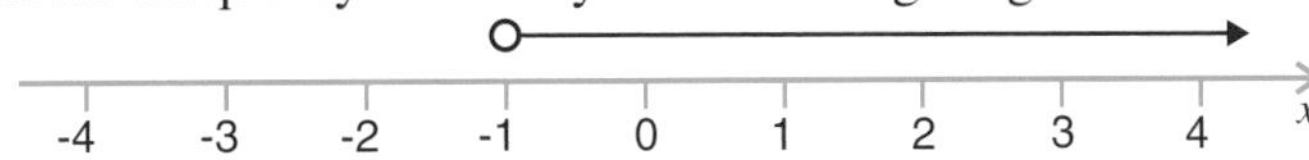

(c) Write down all the integers that satisfy both inequalities shown in parts (a) and (b).

Quadratic Graphs

What you need to know

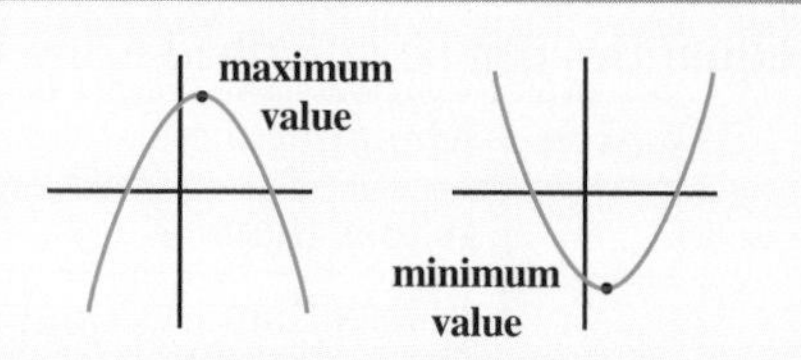

- The graph of a **quadratic function** is a **smooth curve**.
- The general equation for a **quadratic function** is $y = ax^2 + bx + c$, where a cannot be zero.
  The graph of a quadratic function is symmetrical and has a **maximum** or **minimum** value.

You should be able to:
- **substitute** values into given functions to generate points,
- plot graphs of **quadratic functions**,
- use graphs of quadratic functions to solve equations.

Eg 1 (a) Draw the graph of $y = x^2 - 2x - 5$ for values of x from -2 to 4.
(b) Use your graph to solve the equation $x^2 - 2x - 5 = 0$.

(a)

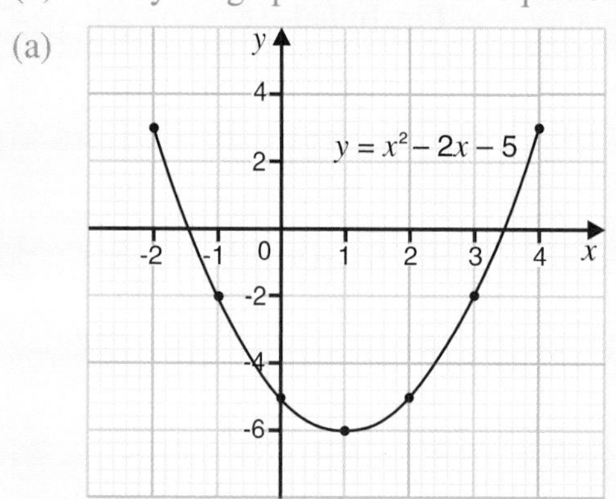

To draw a quadratic graph:
Make a table of values connecting x and y.
Plot the points.
Join the points with a smooth curve.

x	-2	-1	0	1	2	3	4
y	3	-2	-5	-6	-5	-2	3

To solve the equation, read the values of x where the graph of $y = x^2 - 2x - 5$ crosses the x axis ($y = 0$).

(b) $x = -1.4$ and 3.4, correct to one decimal place.

Exercise 20

1 (a) Copy and complete this table of values for $y = x^2 - 2$.

x	-3	-2	-1	0	1	2	3
y		2		-2	-1		7

(b) Draw the graph of $y = x^2 - 2$ for values of x from -3 to 3.
(c) Write down the values of x at the points where the line $y = 3$ crosses your graph.
(d) Write down the values of x where $y = x^2 - 2$ crosses the x axis.

2 (a) Copy and complete the table for the equation $y = x^2 - 3x - 2$.

x	-1	0	1	2	3	4
y			-4	-4	-2	

(b) Hence, draw the graph of $y = x^2 - 3x - 2$.
(c) Use your graph to find the smallest value of y.
(d) Use your graph to find the solutions of the equation $x^2 - 3x - 2 = 0$. OCR

3 (a) Draw the graph of $y = x^2 - 2x + 1$ for values of x from -1 to 3.
(b) Use your graph to find the values of x when $y = 3$. OCR

Algebra Non-calculator Paper

Do not use a calculator for this exercise.

1 Write down the coordinates of the points C and D.

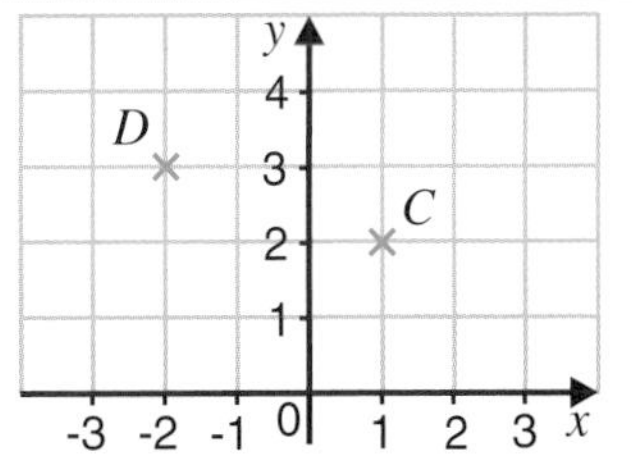

OCR

2 (a) Here is a number pattern. **6, 10, 14, 18, 22, …**
 (i) Write down the next number in this pattern.
 (ii) Explain how you worked out your answer.
(b) Here is a different number pattern. **800, 400, 200, 100, 50, …**
What is the next number? Explain how you worked it out.

OCR

3 In each part, find the output when the input is 12.

(a)
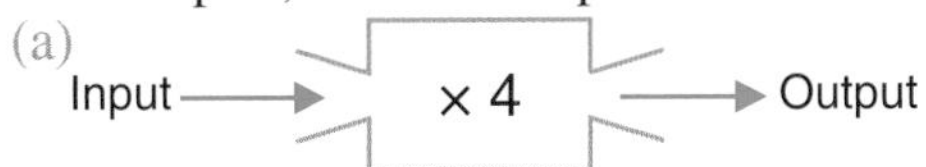

(b)
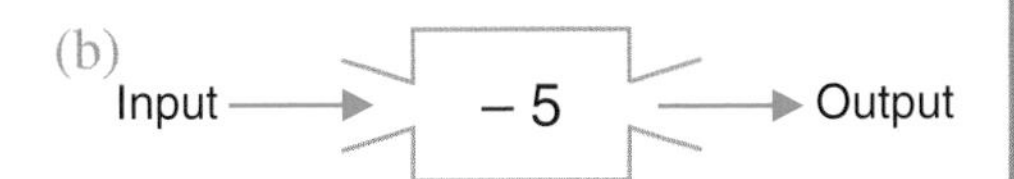

4 Use this rule to find the number of points a football team has scored.

Points scored = 3 × Number of wins + Number of draws

A team wins 7 games and draws 5. How many points have they scored?

5 Regular pentagons are used to form patterns, as shown.

Pattern 1

Pattern 2
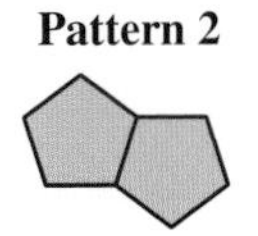

Pattern 3
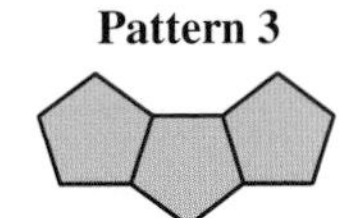

(a) Draw Pattern 4.
(b) Copy and complete the table.

Pattern number	1	2	3	4
Number of sides	5	8	11	

(c) How many sides has Pattern 5?
(d) Pattern 10 has 32 sides. How many sides has Pattern 11?

6 Find the value of $3a + 2b$ when $a = 5$ and $b = 3$.

7 (a) On graph paper, plot the points $A(-3, -2)$ and $B(1, 4)$.
(b) What are the coordinates of the midpoint of the line segment AB?

8 A jam doughnut costs t pence.
(a) Write an expression for the cost of 5 jam doughnuts.

A cream doughnut costs 5 pence more than a jam doughnut.
(b) Write an expression for the cost of a cream doughnut.

9 (a) Find the missing numbers in each of these.
 (i) 16 + ▲ = 30   (ii) ■ + ■ = 10
(b) Here is the start of a number pattern. **37, 35, 33, 31, …**
 (i) What is the next number in the pattern?
 (ii) Explain how you worked out your answer.

OCR

10 $2n$ represents any even number.
Which of the statements describes the number (a) n, (b) $2n + 1$?

always even **always odd** **could be even or odd**

11 Nick thinks of a number. He doubles it and then subtracts 3. The answer is 17.
What is his number?

12 A large envelope costs x pence and a small envelope costs y pence.
Write an expression for the cost of 3 large envelopes and 5 small envelopes.

13 (a) Simplify. $3x + 4y + 7x - 5y$
(b) Multiply out the brackets. $5(7 - 2x)$

OCR

14 Which of these algebraic expressions are equivalent?

$2a - a$	$3a$	$2(a - 1)$	$2a + a$
$2a + 1$	$2a - 2$	$a + a - 1$	2

15 (a) Here are the first four terms of a sequence. 28 26 22 16
Write down the next two terms.
(b) Use the formula $C = \frac{x(y + 2)}{4}$ to work out C when $x = 10$ and $y = 0.4$.

OCR

16 Simplify (a) $7x - 5x + 3x$, (b) $a - 3b + 2a - b$, (c) $3 \times m \times m$.

17 Solve (a) $x + 7 = 4$, (b) $4x = 10$, (c) $\frac{x}{2} = 3$, (d) $2x + 5 = 11$.

18 (a) (i) Copy and complete the table for $y = 3x - 1$.

x	0	1	2	3	4
y					

(ii) Draw the graph of $y = 3x - 1$.
(b) (i) Copy and complete the table for $y = x^2 - 1$.

x	0	1	2	3	4
y		0		8	

(ii) Draw the graph of $y = x^2 - 1$ on the same grid as (a).

OCR

19 Hannah is x years old.
(a) Her sister Louisa is 3 years younger than Hannah.
Write an expression, in terms of x, for Louisa's age.
(b) Their mother is four times as old as Hannah.
Write an expression, in terms of x, for their mother's age.
(c) The total of their ages is 45 years.
By forming an equation in x, find their ages.

20 Given that $S = ut + \frac{1}{2}t^2$, find S when $u = -30$ and $t = 4$.

OCR

21 Ken drives from his home to the city centre.
The graph represents his journey.
(a) How long did Ken take to reach the city centre?
(b) How far from the city centre does Ken live?
(c) What is his average speed for the journey in kilometres per hour?

22 (a) Factorise (i) $3a - 6$, (ii) $k^2 - 2k$.
(b) Cynthia has x five pound notes and $2x$ ten pound notes.
Write an expression, in terms of x, for the total value of her notes.

23 Solve. (a) $2x = 7$ (b) $3x - 5 = 13$ (c) $6x - 9 = x + 26$

OCR

24 Given that $s = 2t^3$, find the value of t when $s = 250$.

25 (a) Expand. $x(x + 2)$
(b) Multiply out the brackets and simplify. $3(4x + 1) + 2(2x - 1)$
(c) (i) Solve. $2x - 5 \geqslant 4$
(ii) Show your solution to part (i) on a number line. OCR

26 (a) $y = \frac{4}{5}(9 - x)$. Find the value of x when $y = 6$.
(b) Solve the equation $\frac{3x + 5}{2} = 7$.

27 (a) On the same diagram draw the graphs $2y = x + 4$ and $y = \frac{1}{2}x + 1$.
(b) What do you notice about the two lines you have drawn?

28 Tickets for a concert cost either £10 or £15.
300 people attended the concert.
x people paid £15 for their ticket.
(a) Write down an expression in x for the amount of money taken for £10 tickets.

The 300 people paid £3950 in total for their tickets.
(b) Write down an equation in x and solve it to find out how many people paid £15. OCR

29 (a) The first five terms of a sequence are 4, 7, 10, 13, 16
Find the nth term of this sequence.
(b) The nth term of another sequence is $2n^2 + 5$.
Find the 10th term of this sequence.

30 (a) Solve the equation. $5(x + 2) = x + 20$
(b) (i) Write down the integer values of n for which $1 < 3n \leqslant 12$.
(ii) Solve the inequality $5x - 2 \geqslant 1$. OCR

31 Match these equations to their graphs.

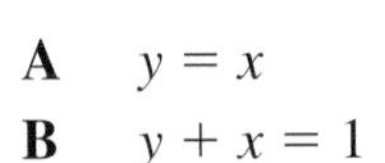

A $y = x$
B $y + x = 1$
C $y = x^2$
D $y = x^3$

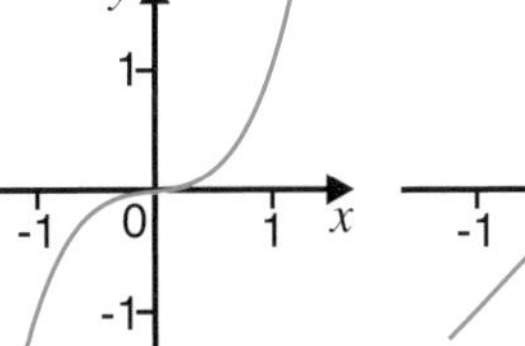

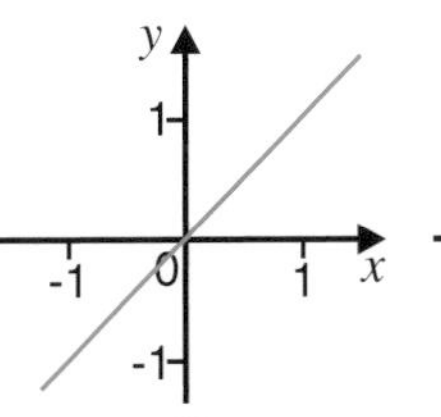

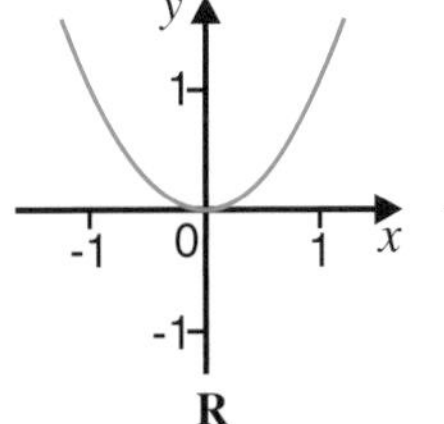

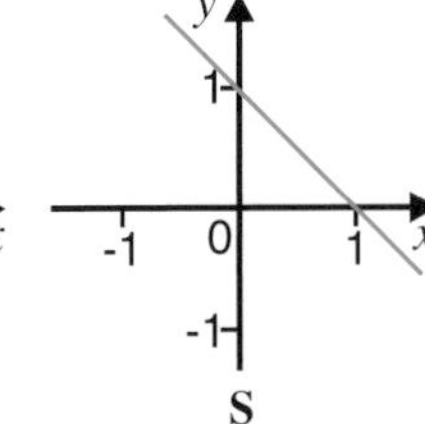

32 (a) Find the first and second term of this sequence. ..., ..., 5, 9, 13, 17
(b) Find a formula for the nth term of this sequence. OCR

33 (a) Copy and complete the table of values for $y = x^2 - 4x + 3$.

x	-1	0	1	2	3	4	5
y			0		0	3	8

(b) Draw the graph of $y = x^2 - 4x + 3$ for values of x from -1 to 5.
(c) Use your graph to solve the equations
(i) $x^2 - 4x + 3 = 0$, (ii) $x^2 - 4x + 3 = 5$.

34 Water is poured into a container at a constant rate.
Copy the axes given and sketch the graph of the depth of the water against time as the container is filled.

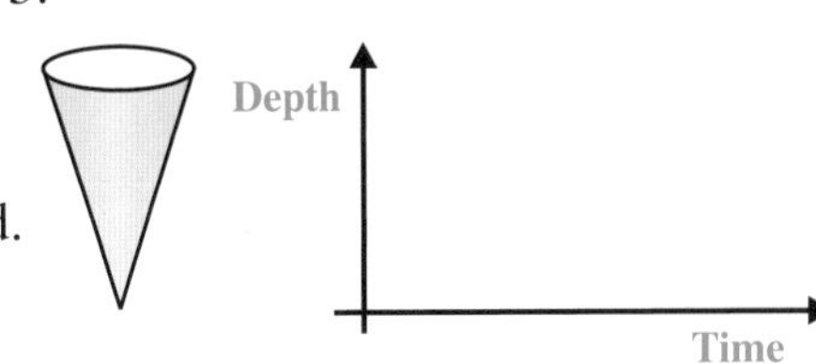

35 (a) Solve $5(x + 1) = 3x + 14$.
(b) Expand and simplify $4(5x - 2) - 2(2x - 1)$. OCR

36 (a) Multiply out and simplify $(p - 2)(p + 2)$.
(b) Simplify $\frac{q^3 \times q^5}{q^2}$.

Algebra Calculator Paper

You may use a calculator for this exercise.

1 (a) Write down the next term in each of the following sequences.

(i) 2, 5, 8, 11, … (ii) 31, 27, 23, 19, … (iii) 3, 4, 6, 9, …

(b) For the sequence in part (a)(ii), describe the rule for finding the next term. OCR

2 (a) Simplify. $9a + 4a - 5a$

(b) Write down, as simply as possible, an expression for the perimeter of this quadrilateral.

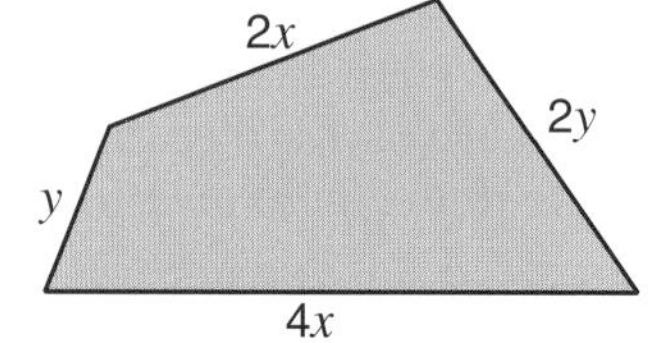

OCR

3 The cost, in £, to hire a bicycle is given by: **Cost = 3 × Number of hours + 2.**

Calculate the cost of hiring two bicycles for 5 hours.

4 Work out the missing values in these calculations.

(a) 

(b) 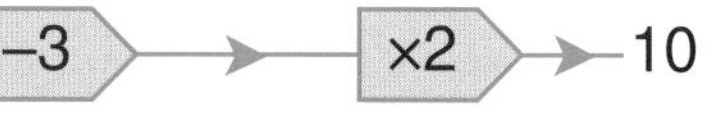

5 (a) A ball costs x pence.
How much will 3 balls cost?

(b) A skipping rope costs 30 pence more than a ball.
How much does a skipping rope cost?

6 (a) (i) What is the next term in this sequence? 2, 9, 16, 23, …

(ii) Will the 50th term in the sequence be an odd number or an even number?
Give a reason for your answer.

(b) Another sequence uses this rule:

Add 3 to the last term.

What term comes before 15 in this sequence?

7 Find the number of crosses in the tenth pattern of this sequence.

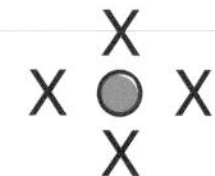

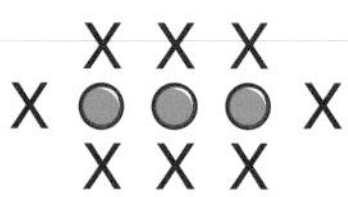

Explain how you worked out your answer. OCR

8 This conversion graph can be used to change euros to dollars.

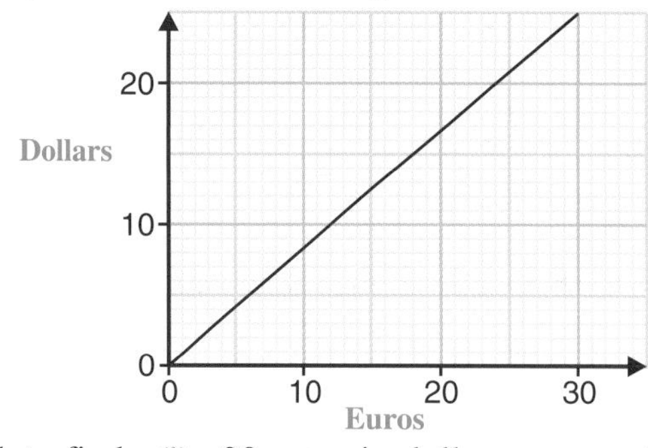

(a) Use the graph to find (i) 30 euros in dollars, (ii) 15 dollars in euros.

(b) Explain how you can use the graph to change 200 dollars into euros.

9 (a) Here is the rule for finding the next term of a sequence.

multiply by 2 then subtract 7

Find the next two terms of the sequence. 5, …, …

(b) Here is a different sequence.

Position	1	2	3	4	5
Term	5	8	11	14	17

(i) Find the 100th term of this sequence.
(ii) Show how you worked out your answer.

OCR

10 (a) Copy and complete the table of values for the equation $y = x - 2$.

x	-1	1	3
y		-1	

(b) Draw the graph of $y = x - 2$ for values of x from -1 to 3.
(c) What are the coordinates of the points where the graph crosses the x axis and the y axis?

11 (a) A shop uses this formula to work out the cost of printing invitations.

multiply 35p by the number of invitations then add £4.50

Teresa orders 40 invitations. How much does she pay?

(b) Another shop uses this formula to work out the cost of printing invitations.
$C = 0.4 \times n$, where C is the cost, in pounds, and n is the number of invitations.
What is the cost of 40 invitations from this shop?

OCR

12 Solve these equations.

(a) $g - 5 = 3$ (b) $4 + a = 9$ (c) $7x = 42$ (d) $5x + 4 = 19$

13 (a) In a week, Phil makes 5 journeys each of length x miles and 8 journeys each of length y miles.
Write down an expression for the number of miles Phil travels in a week.

(b) Find the value of $\frac{3a + 5b}{4}$ when $a = 9$ and $b = -3$.

OCR

14 (a) $T = 3m - 5$. Find the value of m when $T = 4$.
(b) $P = 5y^2$. Find the value of P when $y = 3$.

15 The distance-time graph below shows a journey between Alex's house and Craig's house.

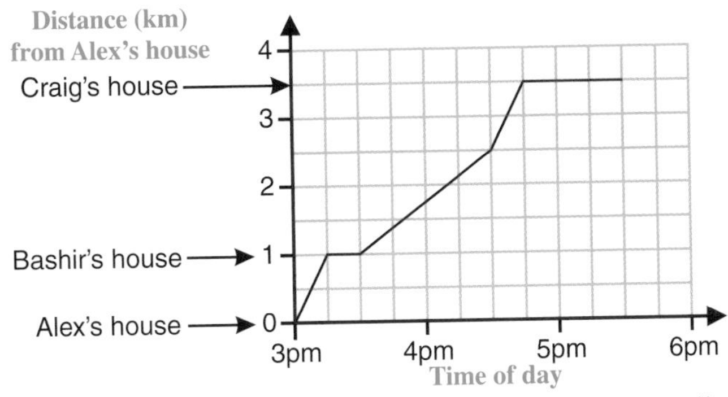

Alex walked from his home to Bashir's house, then they walked together to Craig's house.

(a) How long did Alex spend at Bashir's house?
(b) How far were they from Bashir's house at 4 pm?
(c) What happened at 4.30 pm?
(d) At 5.30 pm, Craig's mother drove Alex home. He arrived home at 5.45 pm. Calculate the average speed of Alex's car journey.

OCR

16 The nth term of a sequence is $2n^2 + 1$.
(a) Find the first term.
(b) Tom says that the fifth term is 101. Explain why he is wrong. OCR

17 (a) Solve the equations (i) $4(a - 2) = 6$, (ii) $5t + 3 = -1 + t$.
(b) The sum of the numbers x, $x - 3$ and $x + 7$ is 25.
By forming an equation in x, find the value of x.

18 Pali's wages each week for his job are made up of two parts.

Basic pay: £227 *Overtime: £9 per hour*

(a) One week, Pali works n hours overtime.
Write down an expression for his total wages this week.
(b) Another week, he was paid £299. Write down an equation in n and solve it to find out how many extra hours he worked that week. OCR

19 (a) Draw the line $y = 2x + 1$ for values of x from -1 to 2.
(b) The line $y = 2x + 1$ crosses the line $x = -5$ at P. Give the coordinates of P.

20 (a) Factorise. (i) $10x + 15$ (ii) $x^2 - 3x$
(b) Solve. $2(5x + 3) = 23$ OCR

21 (a) Write down the next two numbers in this sequence. 26 19 12 5 … …
(b) Here are the first five terms of another sequence. 5 8 11 14 17
Find the nth term of this sequence.

22 The equation $x^3 + 8x - 40 = 0$ has a solution between 2 and 3.
Use trial and improvement to find this solution.
Give your answer correct to two decimal places.
Show clearly the outcomes of your trials. OCR

23 (a) Copy and complete the table of values for $y = x^2 - 5$.

x	-3	-2	-1	0	1	2	3
y	4		-4	-5			4

(b) Draw the graph of $y = x^2 - 5$ for values of x from -3 to 3.
(c) Use your graph to solve the equation $x^2 - 5 = 0$.

24 (a) Multiply out the brackets and simplify this expression. $3(2x + 3) + 2(4x - 1)$
(b) Rearrange the formula $P = 2L + 2W$ to make L the subject. OCR

25 A glass of milk costs x pence. A milk shake costs 45 pence more than a glass of milk.
(a) Write an expression for the cost of a milk shake.
(b) Lou has to pay £4.55 for 3 milk shakes and a glass of milk.
By forming an equation, find the price of a glass of milk.

26 (a) Solve the inequality $3x - 5 > 4$.
(b) List the values of x, where x is an integer, such that $-1 \leqslant x + 2 < 1$.

27 (a) Simplify. $a^3 \times a^4$
(b) Rearrange the formula $P = 3t + 5$ to find a formula for t in terms of P. OCR

28 Use a trial and improvement method to find the value of x, correct to **one** decimal place, when $x^3 - 2x = 68$. You must show all your trials. OCR

29 (a) Solve the equation $3 - x = 4(x + 1)$.
(b) Multiply out and simplify $2(5x - 3) - 3(x - 1)$.
(c) Simplify (i) $m^8 \div m^2$, (ii) $n^2 \times n^3$.

SECTION 21

Angles

What you need to know

- You should be able to use a **protractor** to measure and draw angles accurately.

Eg 1 Measure the size of this angle.

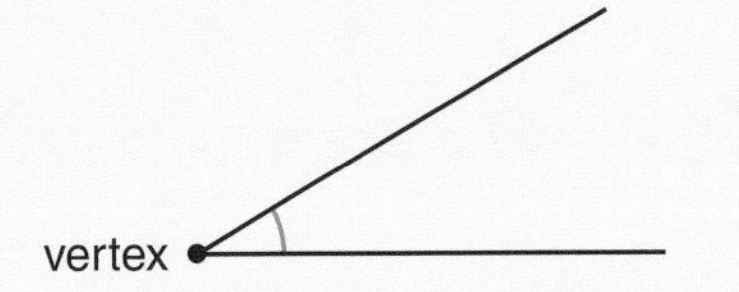

The angle measures 30°.

To measure an angle, the protractor is placed so that its centre point is on the corner (vertex) of the angle, with the base along one of the arms of the angle, as shown.

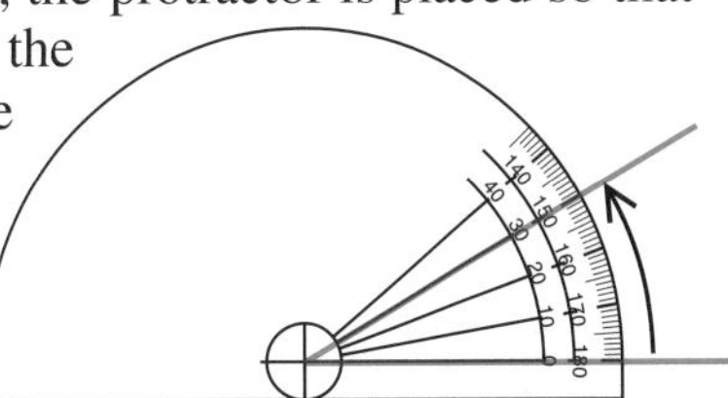

- Types and names of angles.

Acute angle

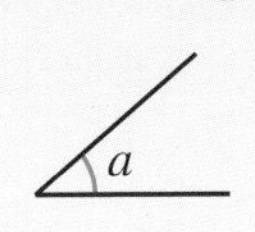

$0° < a < 90°$

Right angle

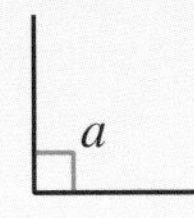

$a = 90°$

Obtuse angle

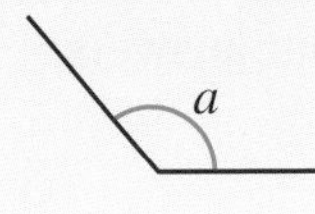

$90° < a < 180°$

Reflex angle

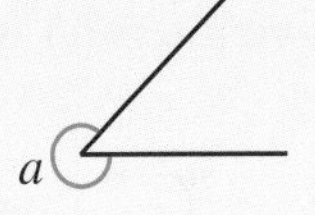

$180° < a < 360°$

- Angle properties.

Angles at a point

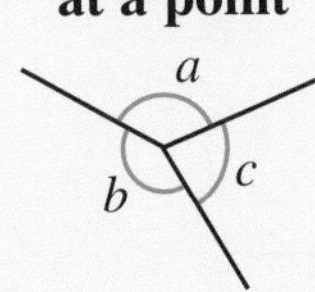

$a + b + c = 360°$

Complementary angles

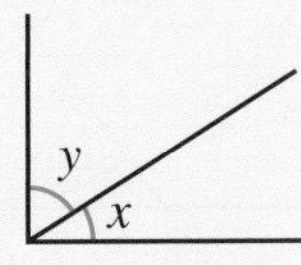

$x + y = 90°$

Supplementary angles

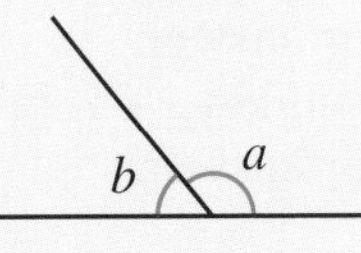

$a + b = 180°$

Vertically opposite angles

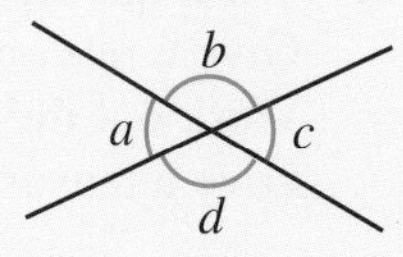

$a = c$ and $b = d$

- A straight line joining two points is called a **line segment**.
- Lines which meet at right angles are **perpendicular** to each other.
- Lines which never meet and are always the same distance apart are **parallel**.
- When two parallel lines are crossed by a **transversal** the following pairs of angles are formed.

Corresponding angles

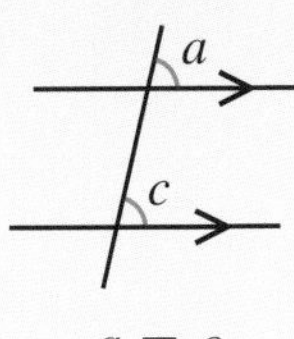

$a = c$

Alternate angles

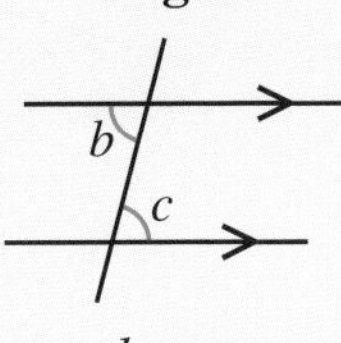
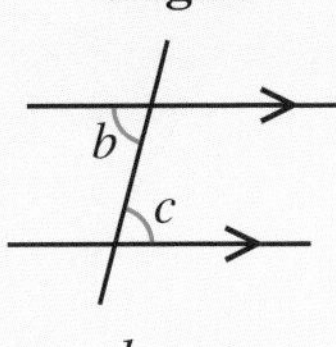

$b = c$

Allied angles

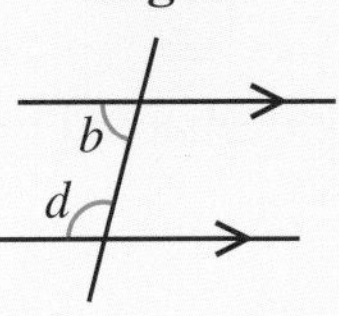

$b + d = 180°$

Arrowheads are used to show that lines are **parallel**.

- You should be able to use angle properties to solve problems involving lines and angles.

Eg 2 Work out the size of the angles marked with letters.
Give a reason for each answer.

$a + 64° = 180°$ (supplementary angles)
$a = 180° - 64° = 116°$

$b = 64°$ (vertically opposite angles)
$c = 64°$ (corresponding angles)

Exercise 21

1

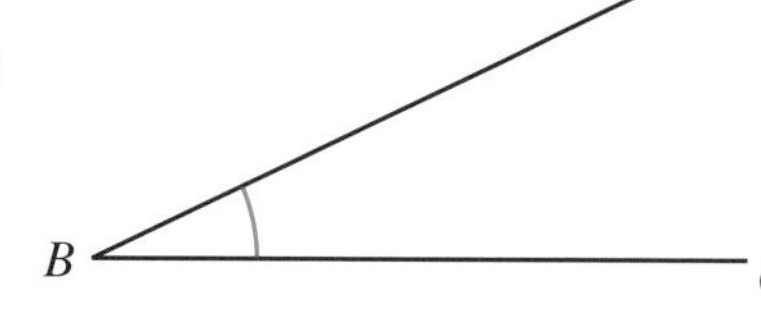

(a) Measure the size of angle ABC.
(b) What special name is given to angle ABC?

2 Look at the diagram.
(a) Which lines are parallel to each other?
(b) Which lines are perpendicular to each other?
(c) (i) Measure angle y.
(ii) Which of these words describes angle y?

acute angle **obtuse angle** **reflex angle**

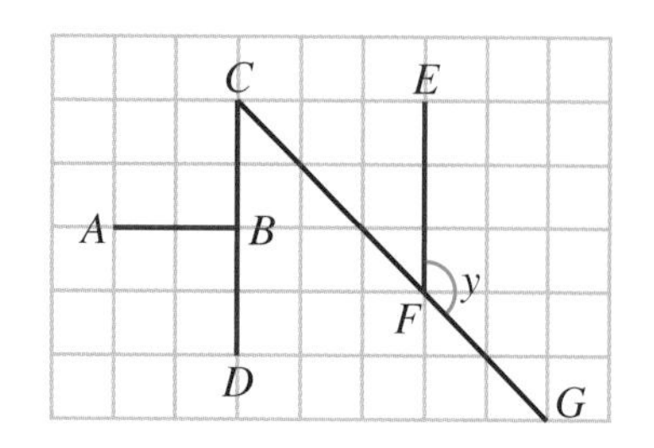

3

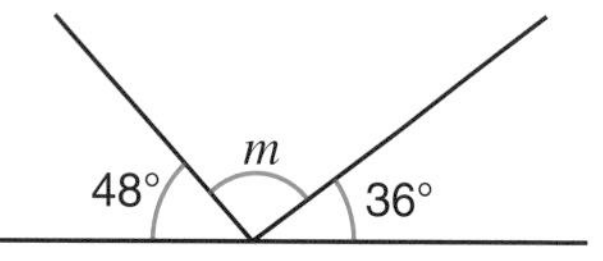

(a) Draw an angle of 112°.
(b) Find angle m.
Give a reason for your answer.

OCR

4 Without measuring, find the size of the lettered angles.
Give a reason for each of your answers.

(a)

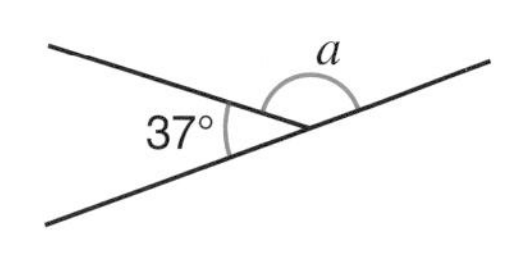

(b)

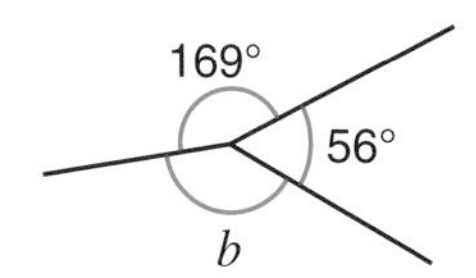

(c)

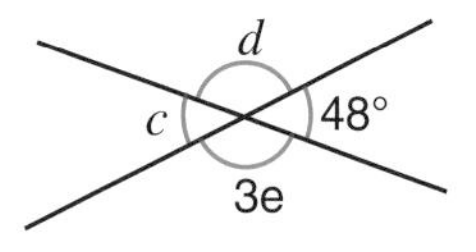

5 AB is parallel to DC.
(a) Work out the size of angle x.
Give a reason for your answer.
(b) Work out the size of angle y.
Give a reason for your answer.

6

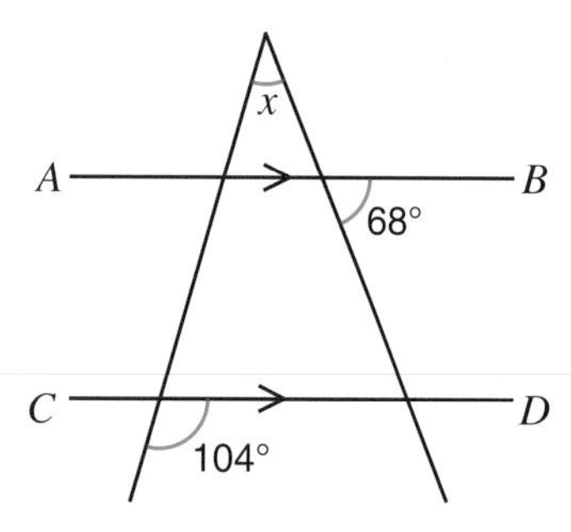

In the diagram, AB is parallel to CD.
Find the value of x.

OCR

7 Work out the size of the angles marked with letters.
Give a reason for each answer.

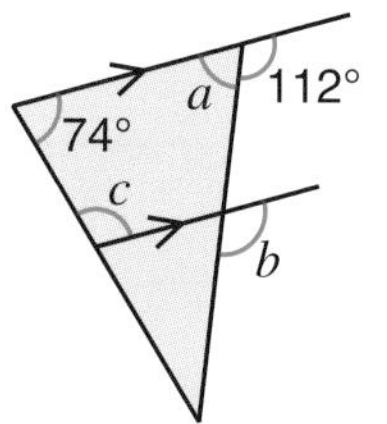

8 Find the size of the angles marked with letters.

(a)

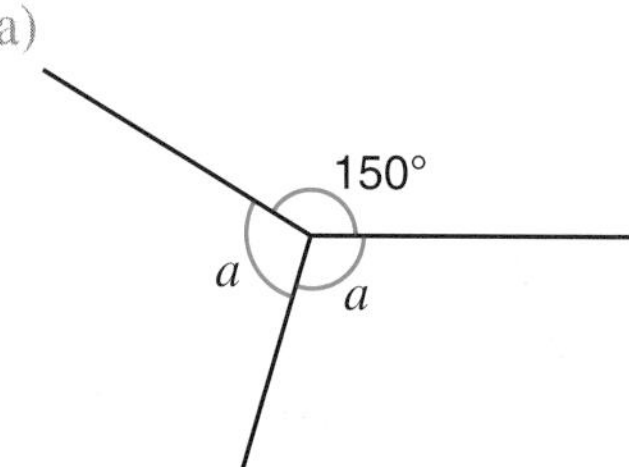

(b)

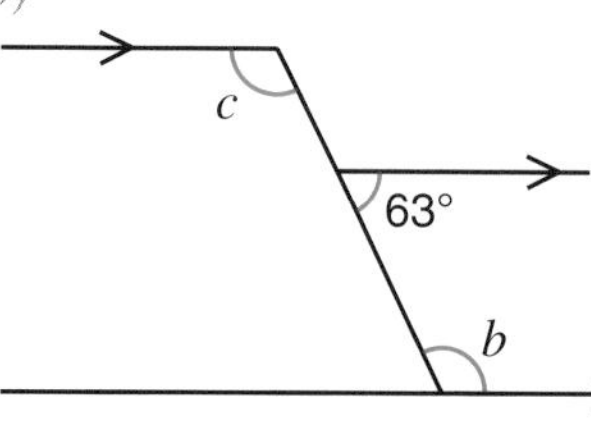

(c)

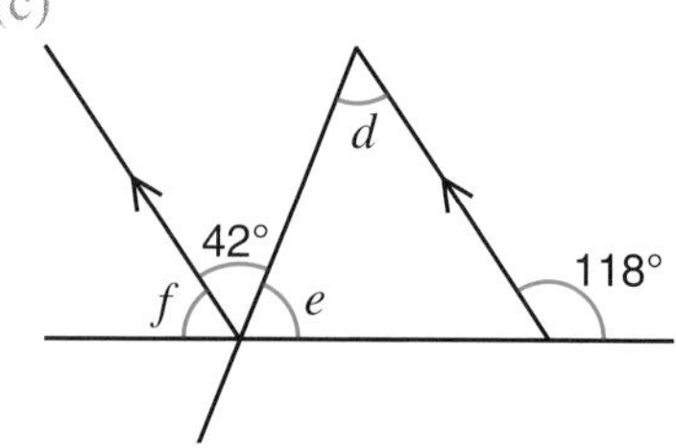

Triangles

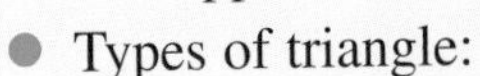

- Triangles can be: **acute-angled** (all angles less than 90°),
  obtuse-angled (one angle greater than 90°),
  right-angled (one angle equal to 90°).
- The sum of the angles in a triangle is 180°.
  $a + b + c = 180°$
- The exterior angle is equal to the sum of the two opposite interior angles. $a + b = d$

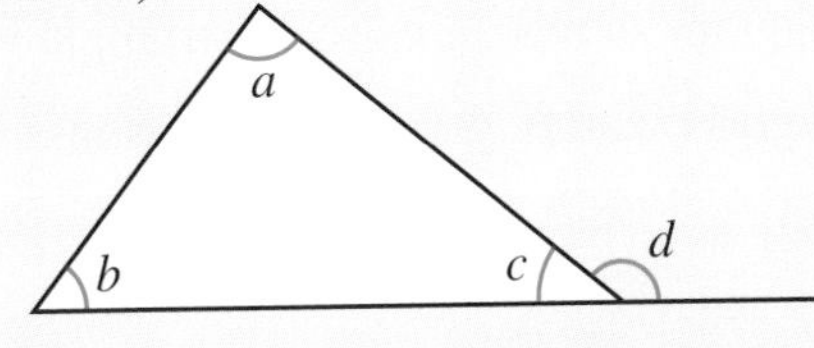

- Types of triangle:

Scalene

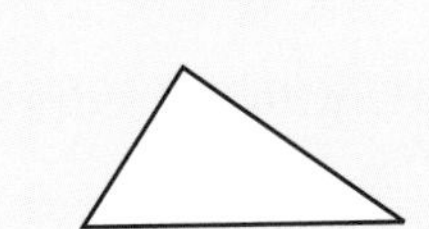

Sides have different lengths.
Angles are all different.

Isosceles

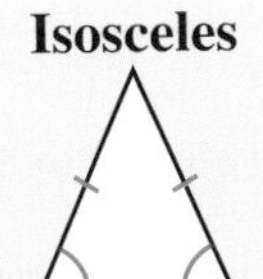

Two equal sides.
Two equal angles.

Equilateral

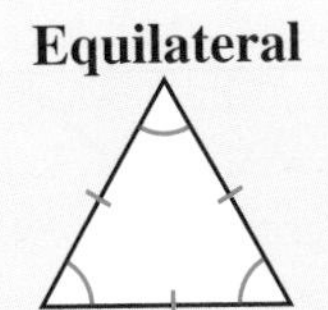

Three equal sides.
Three equal angles, 60°.

> A **sketch** is used when an accurate drawing is not required.
> Dashes across lines show sides that are equal in length.
> Equal angles are marked using arcs.

- You should be able to use properties of triangles to solve problems.

Eg 1 Find the size of the angles marked a and b.

$a = 86° + 51°$ (ext. ∠ of a Δ)
$a = 137°$
$b + 137° = 180°$ (supp. ∠'s)
$b = 43°$

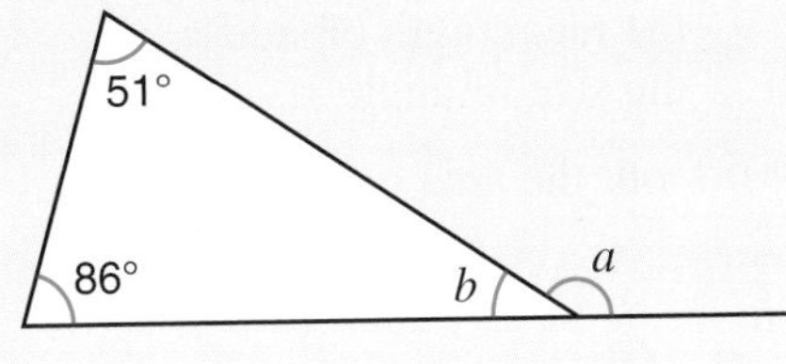

- Perimeter of a triangle is the sum of its three sides.
- Area of a triangle $= \frac{\text{base} \times \text{perpendicular height}}{2}$

$A = \frac{1}{2} \times b \times h$

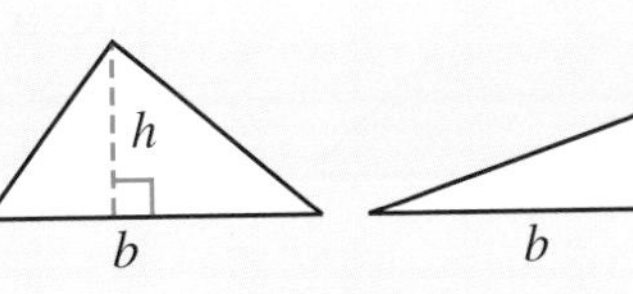

Eg 2 Calculate the area of this triangle.

$A = \frac{1}{2} \times b \times h$
$= \frac{1}{2} \times 9 \times 6\,\text{cm}^2$
$= 27\,\text{cm}^2$

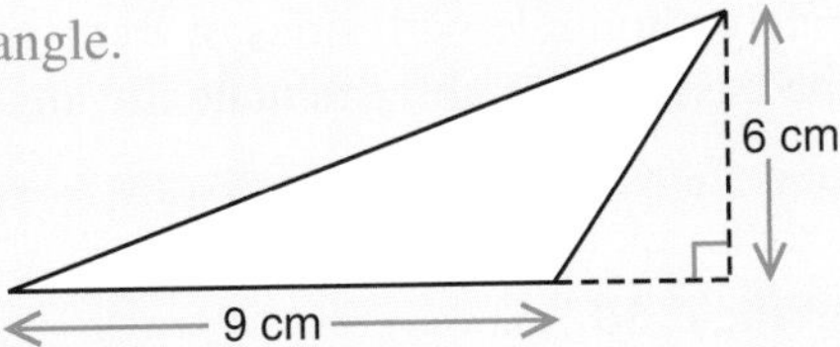

- You should be able to draw triangles accurately, using ruler, compasses and protractor.

Exercise 22

1. Without measuring, work out the size of the angles marked with letters.

(a)

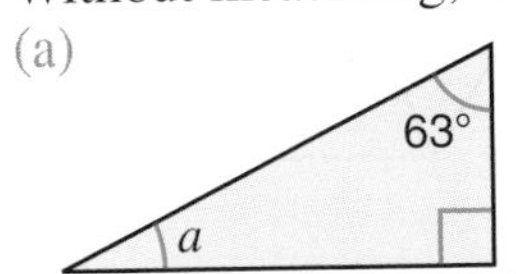

(b)

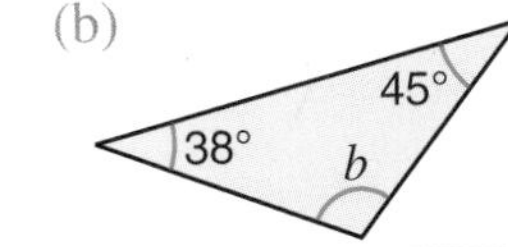

(c)

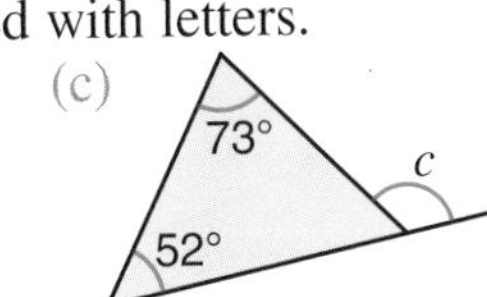

2 Work out angle a and angle b.

Give a reason for each answer.

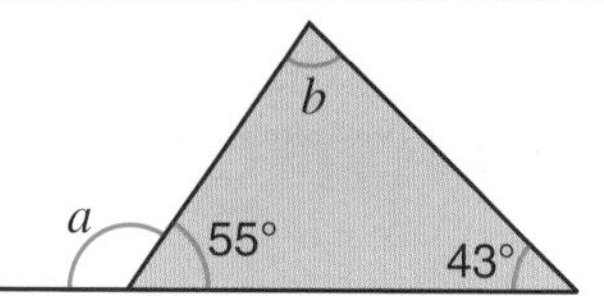

OCR

3 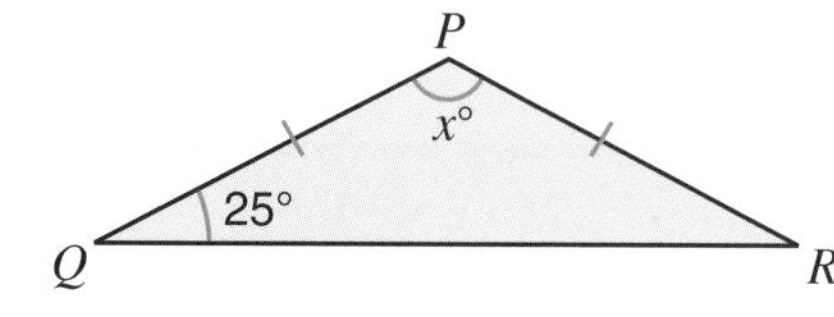

The diagram shows triangle PQR, with $PQ = PR$.

Work out the value of x.
Give a reason for your answer.

4 This diagram shows an isosceles triangle with two of its sides extended.

Find angles a and b, giving your reasons.

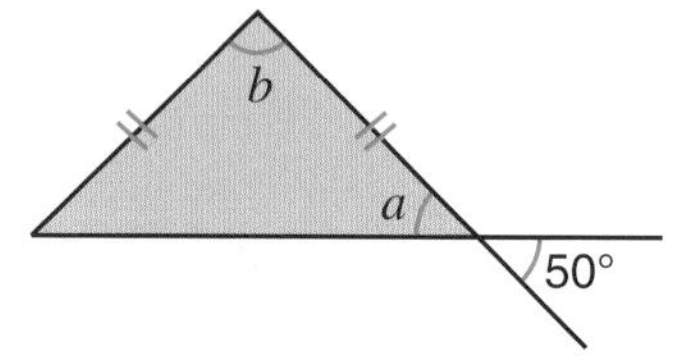

OCR

5 The diagram shows an isosceles triangle with two sides extended.

(a) Work out the size of angle x.

(b) Work out the size of angle y.

6 (a) Make an accurate scale drawing of this triangle on one centimetre squared paper.
Use a scale of 1 cm to 2 cm.

(b) Use your drawing to find
(i) the real length of side a,
(ii) the size of angle x.

(c) Work out the area of this triangle.

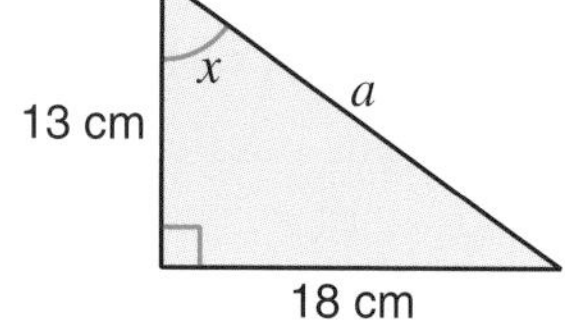

OCR

7 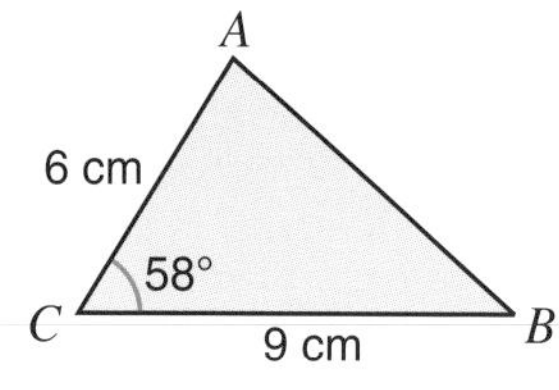

Make a full-size drawing of this triangle.

OCR

8 (a) Construct accurately a triangle with sides of 8 cm, 6 cm and 5 cm.
(b) By measuring the base and height, calculate the area of the triangle.

9 The diagram shows triangle ABC.

Calculate the area of triangle ABC.

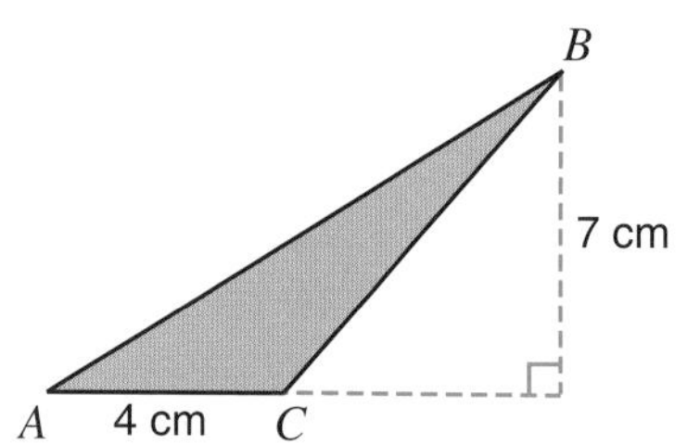

10 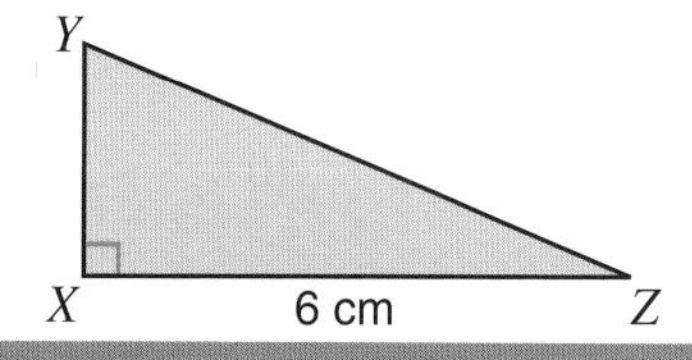

Triangle XYZ has an area of $12\,\text{cm}^2$.
$XZ = 6$ cm.

Calculate YX.

Symmetry and Congruence

What you need to know

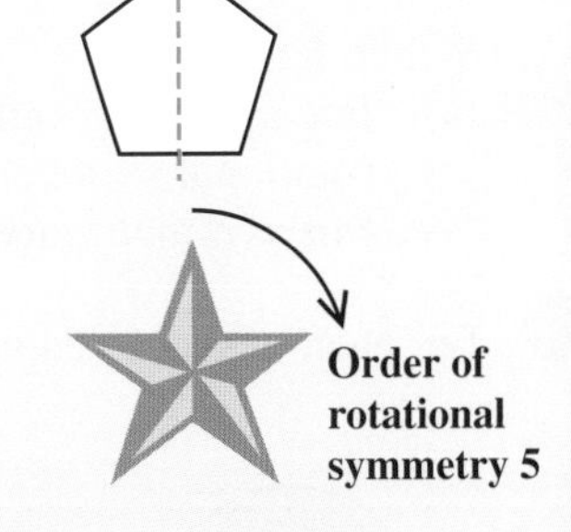

- A two-dimensional shape has **line symmetry** if the line divides the shape so that one side fits exactly over the other.
- A two-dimensional shape has **rotational symmetry** if it fits into a copy of its outline as it is rotated through 360°.
- A shape is only described as having rotational symmetry if the order of rotational symmetry is 2 or more.
- The number of times a shape fits into its outline in a single turn is the **order of rotational symmetry**.

Eg 1 For each of these shapes (a) draw and state the number of lines of symmetry,
(b) state the order of rotational symmetry.

(i)

Two lines of symmetry.
Rotational symmetry of order 2.

(ii)

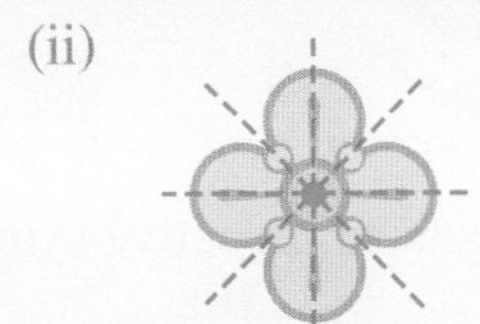

4 lines of symmetry.
Order of rotational symmetry 4.

(iii)

No lines of symmetry.
Order of rotational symmetry 1.
The shape is **not** described as having rotational symmetry.

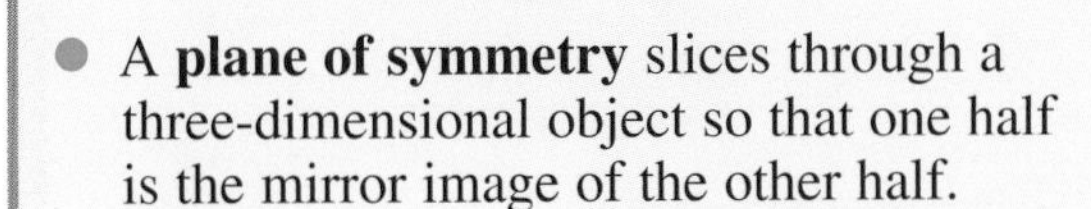

- A **plane of symmetry** slices through a three-dimensional object so that one half is the mirror image of the other half.

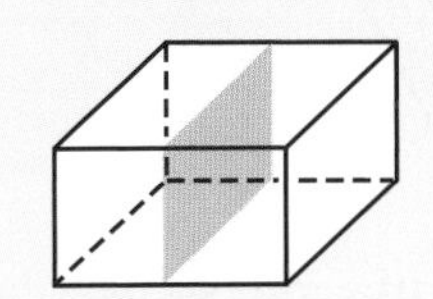

- Three-dimensional objects can have **axes of symmetry**.

Eg 2 Sketch a cuboid and show its axes of symmetry.

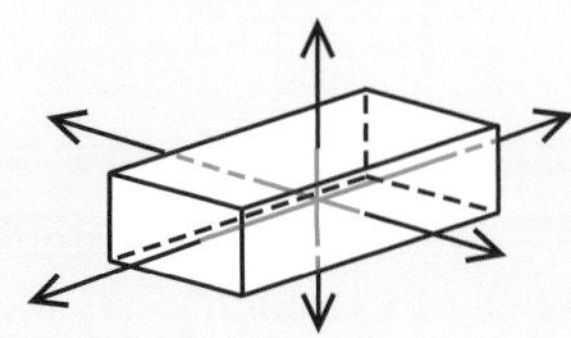

A cuboid has three axes of symmetry.
The order of rotational symmetry about each axis is 2.

- When two shapes are the same shape and size they are said to be **congruent**.
- There are four ways to show that a pair of triangles are congruent.

SSS	3 corresponding sides.	**ASA**	2 angles and a corresponding side.
SAS	2 sides and the included angle.	**RHS**	Right angle, hypotenuse and one other side.

Eg 3 Which of these triangles are congruent to each other? Give a reason for your answer.

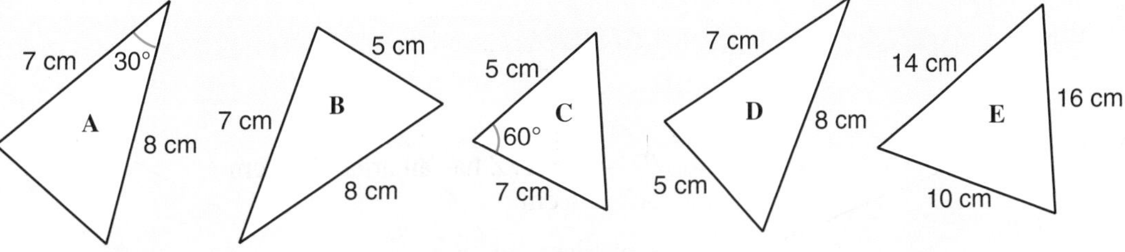

B and D. Reason: 3 corresponding sides (SSS)

Exercise 23

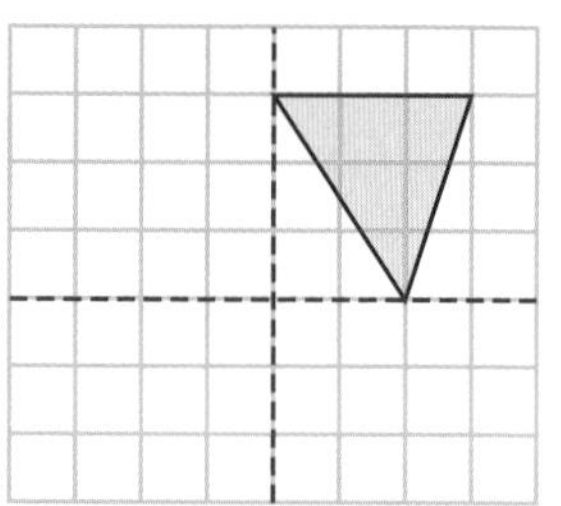

1 The diagram shows part of a design.
The dotted lines are lines of symmetry of the whole design.

(a) Copy the complete the design.

(b) Write down the order of the rotational symmetry of the completed design.

OCR

2 Consider the letters of the word **O R A N G E**
Which letters have
(a) line symmetry only,
(b) rotational symmetry only,
(c) line symmetry and rotational symmetry?

3 For each of these shapes state (i) the number of lines of symmetry,
(ii) the order of rotational symmetry.

(a)

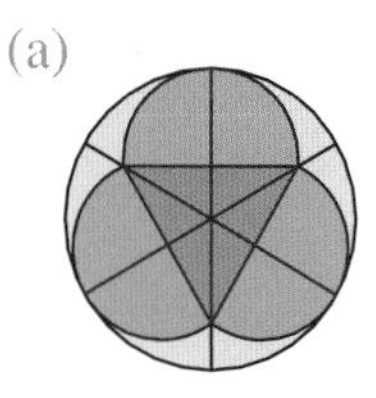

(b)

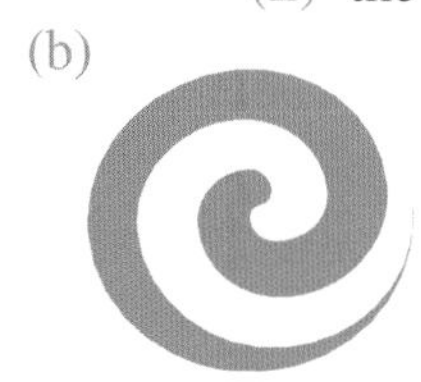

(c)

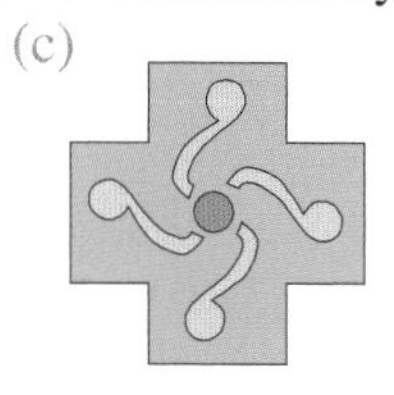

(d)

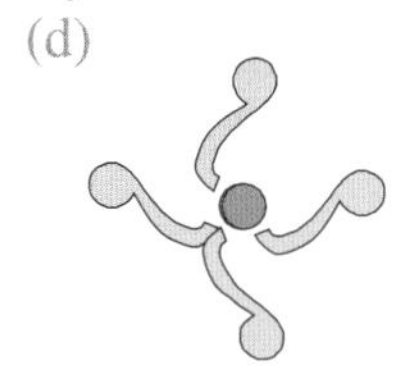

4 (a) Which of these kitchen items have reflective symmetry?

(i) **Tray**

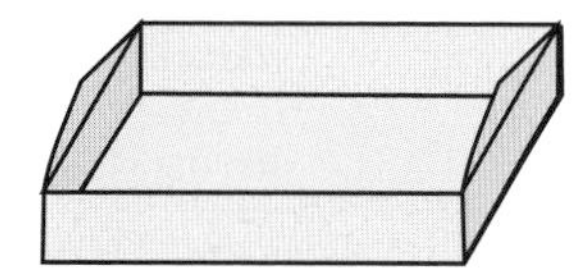

(ii) **Jug**

(iii) **Saucepan**

(b) Shade 4 more squares of this grid so that the dashed line is the **only** line of symmetry.

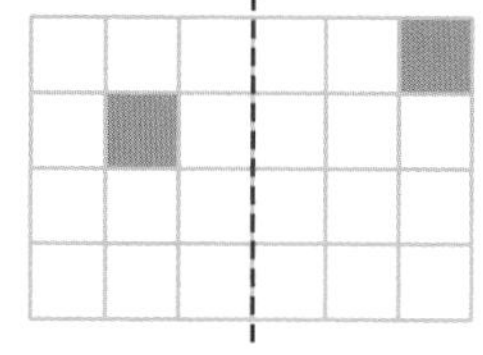

OCR

5 The diagram shows a square-based pyramid.
(a) How many planes of symmetry has the pyramid?
(b) How many axes of symmetry has the pyramid?

6 The diagram shows a rectangle which has been cut into 6 pieces.
Which two pieces are congruent to each other?

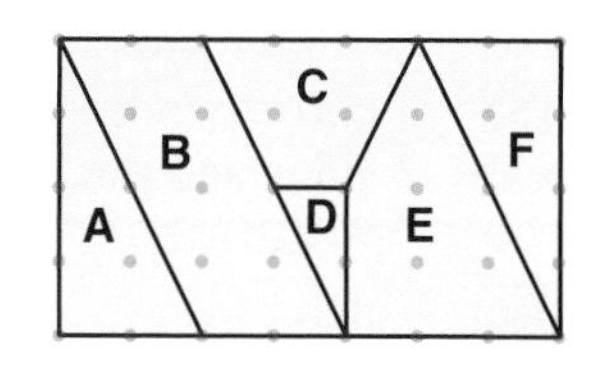

7 The diagram shows information about four triangles.
Which two triangles are congruent? Give a reason for your answer.

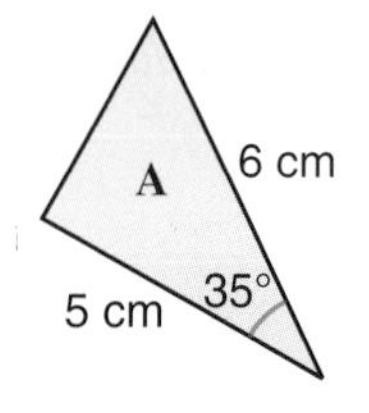

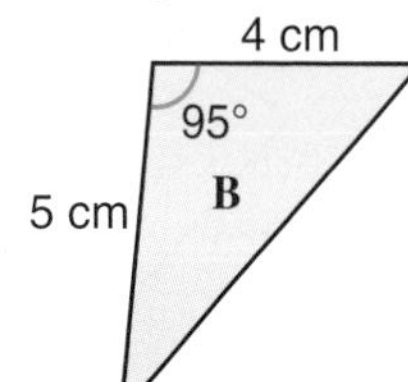

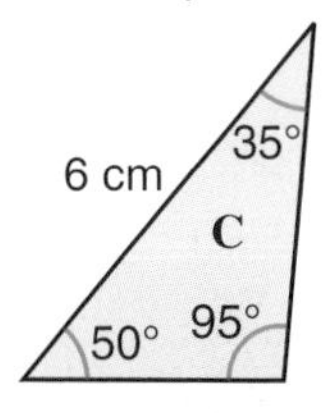

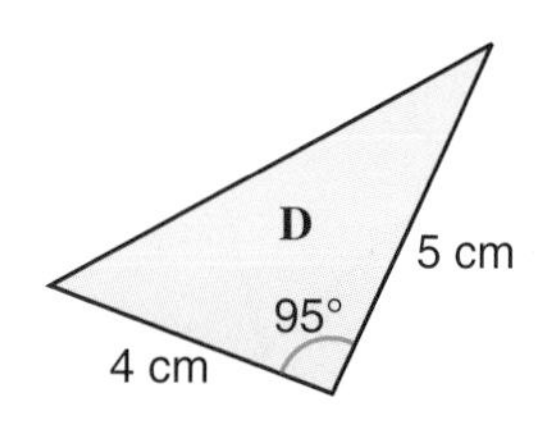

Quadrilaterals

What you need to know

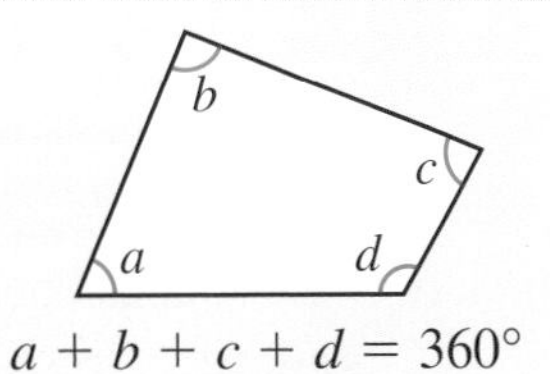

$a + b + c + d = 360°$

- A **quadrilateral** is a shape made by four straight lines.
- The sum of the angles in a quadrilateral is 360°.
- The **perimeter** of a quadrilateral is the sum of the lengths of its four sides.
- Facts about these special quadrilaterals:

rectangle **square** **parallelogram** **rhombus** **trapezium** **isosceles trapezium** **kite**

Quadrilateral	Sides	Angles	Diagonals	Line symmetry	Order of rotational symmetry	Area formula
Rectangle	Opposite sides equal and parallel	All 90°	Bisect each other	2	2	A = length × breadth $A = lb$
Square	4 equal sides, opposite sides parallel	All 90°	Bisect each other at 90°	4	4	A = (length)^2 $A = l^2$
Parallelogram	Opposite sides equal and parallel	Opposite angles equal	Bisect each other	0	2	A = base × height $A = bh$
Rhombus	4 equal sides, opposite sides parallel	Opposite angles equal	Bisect each other at 90°	2	2	A = base × height $A = bh$
Trapezium	1 pair of parallel sides					$A = \frac{1}{2}(a + b)h$
Isosceles trapezium	1 pair of parallel sides, non-parallel sides equal	2 pairs of equal angles	Equal in length	1	1*	$A = \frac{1}{2}(a + b)h$
Kite	2 pairs of adjacent sides equal	1 pair of opposite angles equal	One bisects the other at 90°	1	1*	

*A shape is only described as having rotational symmetry if the order of rotational symmetry is 2 or more.

- You should be able to use properties of quadrilaterals to solve problems.

Eg 1 Work out the size of the angle marked x.

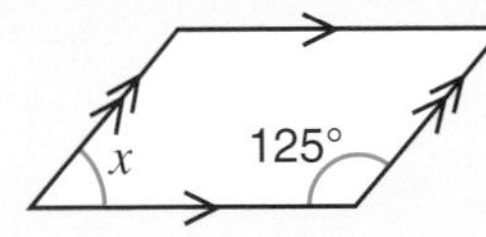

Opposite angles are equal.
So, $125° + 125° + x + x = 360°$
$x = 55°$

Eg 2 Find the area of this trapezium.

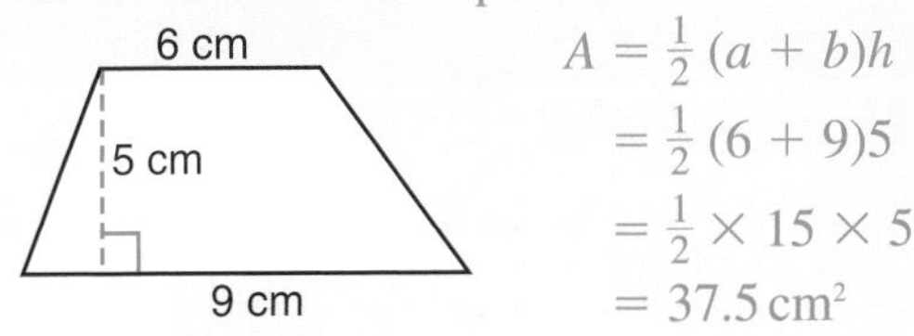

$A = \frac{1}{2}(a + b)h$
$= \frac{1}{2}(6 + 9)5$
$= \frac{1}{2} \times 15 \times 5$
$= 37.5\ \text{cm}^2$

- You should be able to construct a quadrilateral from given information using ruler, protractor, compasses.

Exercise 24

1 This rectangle is drawn on 1 cm squared paper.
It has a perimeter of 18 cm.

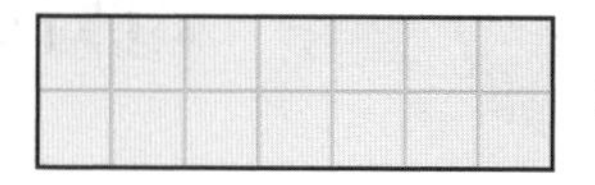

Not full size

(a) What is the area of the rectangle?
(b) (i) On 1 cm squared paper draw three different rectangles which each have a perimeter of 18 cm.
(ii) Find the area of each rectangle.

2 (a) Here is a description of a quadrilateral.

The diagonals bisect each other and are at 90°.

Which of these shapes could it be?

square **rectangle** **rhombus** **kite** **trapezium**

(b) Here are two statements which are true for all quadrilaterals.
- They have four sides.
- They have four angles.

Write down another mathematical statement which is true for parallelograms, but is not true for **all** quadrilaterals.

OCR

3 Find the size of the lettered angles.

(a)

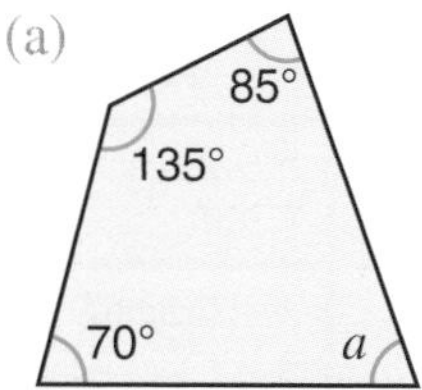

(b)

48°
b

(c)

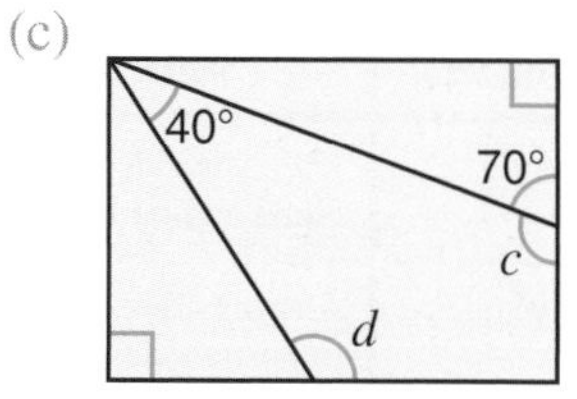

4 The diagram shows a quadrilateral $ABCD$.
$AB = BC$ and $CD = DA$.

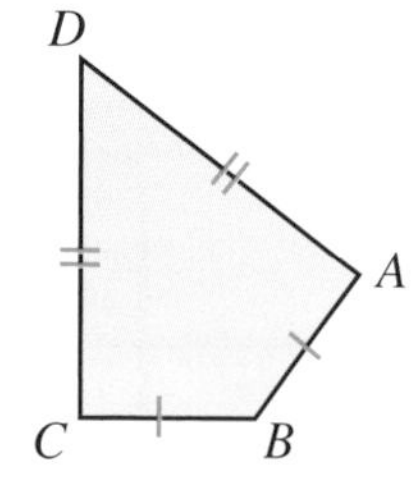

(a) Which of the following correctly describes the quadrilateral $ABCD$?

rhombus **parallelogram** **kite** **trapezium**

(b) Angle $ADC = 36°$ and angle $BCD = 105°$.
Work out the size of angle ABC.

5 Aisha buys a rectangular carpet measuring 5 m by 4 m. The carpet costs £14 per square metre.
How much does she pay?

OCR

6

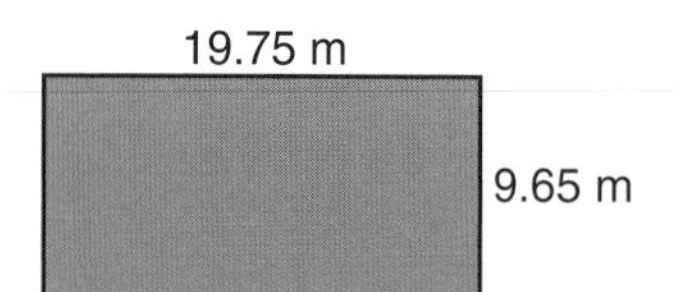

The diagram shows a rectangular garden.
Selina wants to estimate the area of the garden.
(a) Write down a calculation she could do in her head to estimate the area of the garden.
(b) Is your estimate bigger or smaller than the exact area?
Explain how you decide.

OCR

7 (a) In the diagram, CDE is a straight line,
AB is parallel to CD and $BD = DE$.
Angle $BDE = 110°$.
Angle BAC and angle ACD are right angles.
(i) Find the value of x.
Give a reason for your answer.
(ii) Find the value of y and z.

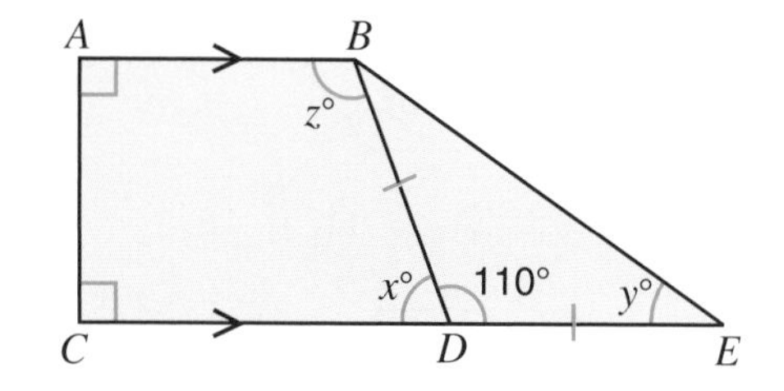

(b)

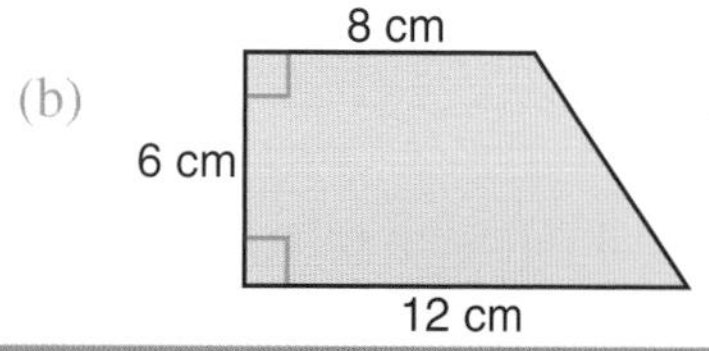

Find the area of this trapezium.

OCR

SECTION 25

Polygons

What you need to know

- A **polygon** is a many-sided shape made by straight lines.
- A polygon with all sides equal and all angles equal is called a **regular polygon**.
- Shapes you need to know:
  A 3-sided polygon is called a **triangle**.
  A 4-sided polygon is called a **quadrilateral**.
  A 5-sided polygon is called a **pentagon**.
  A 6-sided polygon is called a **hexagon**.
  An 8-sided polygon is called an **octagon**.
- The sum of the exterior angles of any polygon is 360°.
- At each vertex of a polygon: interior angle + exterior angle = 180°
- The sum of the interior angles of an n-sided polygon is given by:
  $(n - 2) \times 180°$
- For a regular n-sided polygon: exterior angle $= \frac{360°}{n}$
- You should be able to use the properties of polygons to solve problems.

interior angle
exterior angle

Eg 1 Find the sum of the interior angles of a pentagon.
$(5 - 2) \times 180° = 3 \times 180° = 540°$

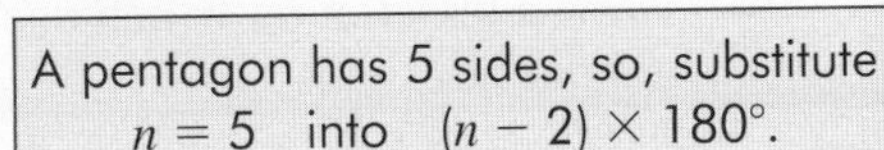
A pentagon has 5 sides, so, substitute $n = 5$ into $(n - 2) \times 180°$.

Eg 2 A regular polygon has an exterior angle of 30°.
(a) How many sides has the polygon?
(b) What is the size of an interior angle of the polygon?

(a) $n = \frac{360°}{\text{exterior angle}}$

$n = \frac{360°}{30°}$

$n = 12$

(b) int. ∠ + ext. ∠ = 180°
int. ∠ + 30° = 180°
interior angle = 150°

- A shape will **tessellate** if it covers a surface without overlapping and leaves no gaps.

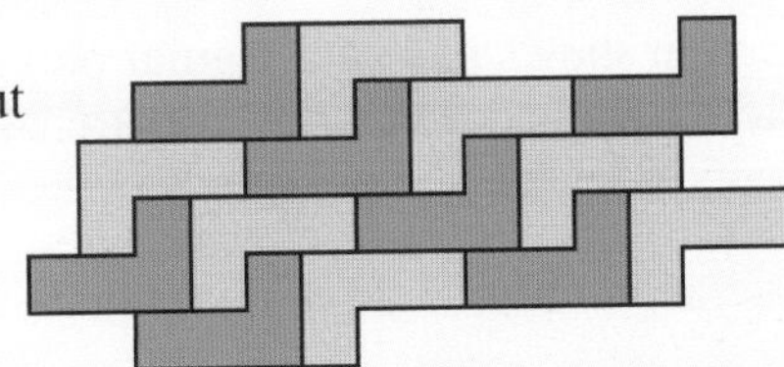

- All triangles tessellate.
- All quadrilaterals tessellate.
- Equilateral triangles, squares and hexagons can be used to make **regular tessellations**.

Exercise 25

1. These shapes are drawn on isometric paper.

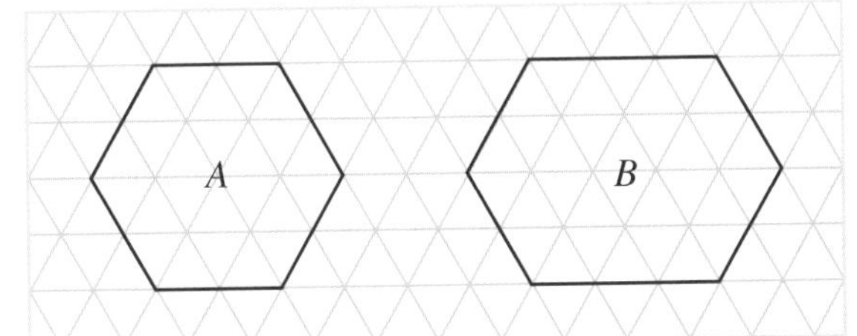

What are the differences between the symmetry of shape A and the symmetry of shape B?

2 $ABCDE$ is a regular pentagon.

Calculate the size of angle x.
Give reasons for your answer.

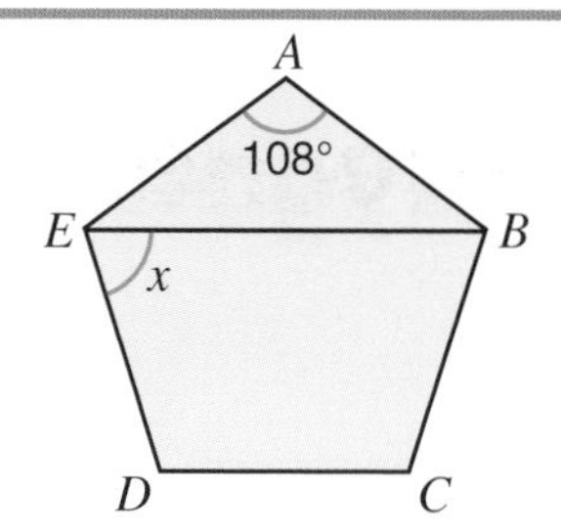

OCR

3 Work out the size of the angles marked with letters.

(a)

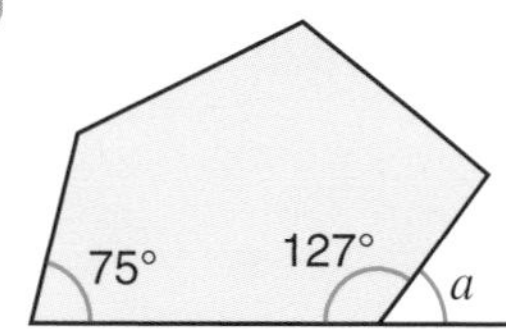

(b)

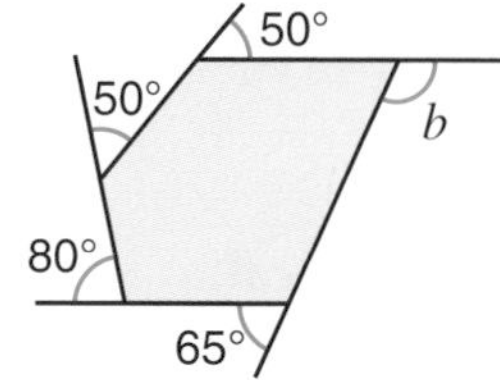

(c)

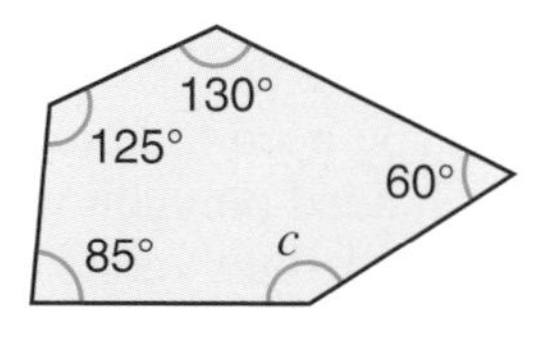

4 These shapes are regular polygons. Work out the size of the lettered angles.

(a)

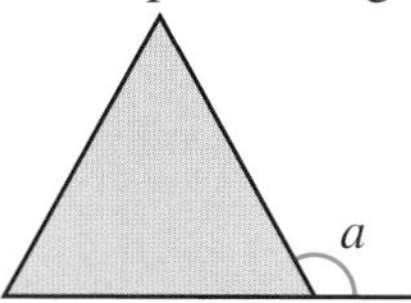

(b)

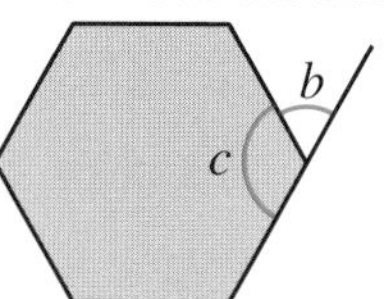

(c)

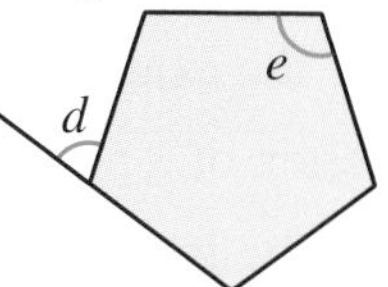

5 (a) Here is a pattern of regular octagons and squares. Explain why these shapes tessellate.

(b) Draw a tessellation which uses equilateral triangles and regular hexagons.

6

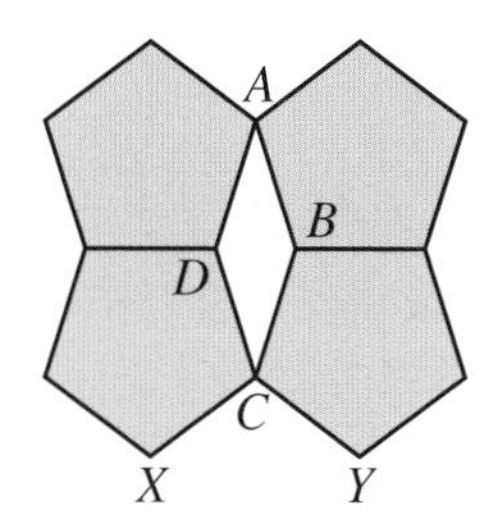

Four regular pentagons are placed together, as shown, to form a rhombus, $ABCD$.

Calculate the size of

(a) angle ABC,

(b) angle XCY.

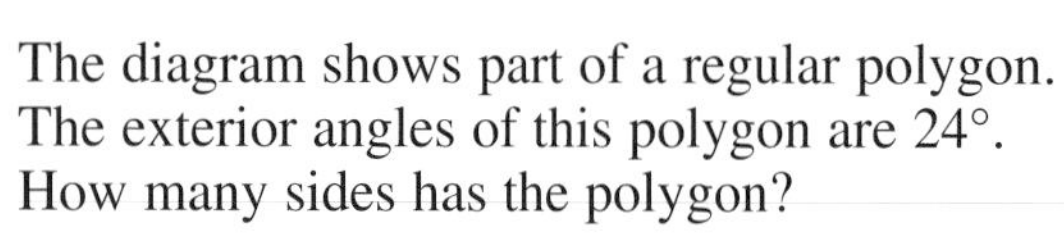

7 The diagram shows part of a regular polygon.
The exterior angles of this polygon are 24°.
How many sides has the polygon?

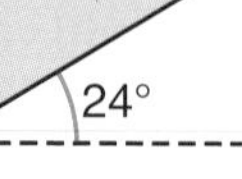

8 Calculate an interior angle of a regular 9-sided polygon.

OCR

9

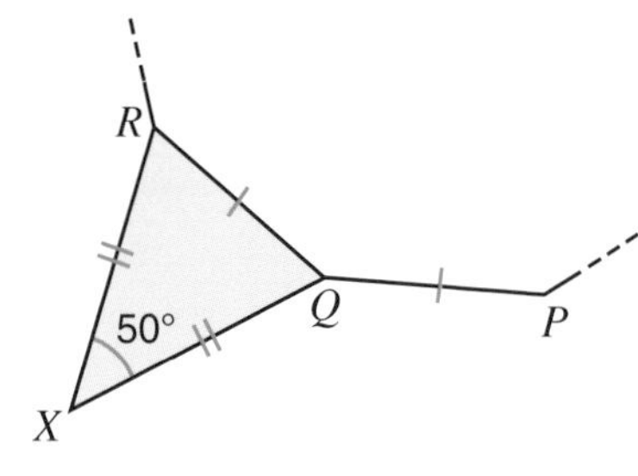

PQ and QR are two sides of a regular 10-sided polygon.
QRX is an isosceles triangle with $RX = XQ$.
Angle $QXR = 50°$.

Work out the size of the obtuse angle PQX.

10 The diagram shows part of an inscribed regular polygon.
The line AB is one side of the polygon.
O is the centre of the circle.
Angle $AOB = 30°$.

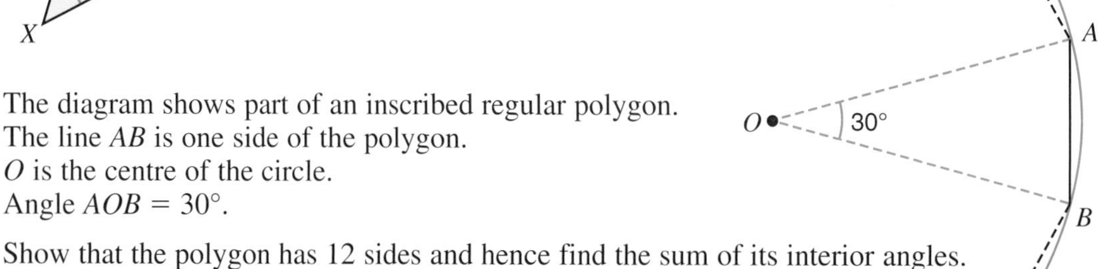

Show that the polygon has 12 sides and hence find the sum of its interior angles.

Direction and Distance

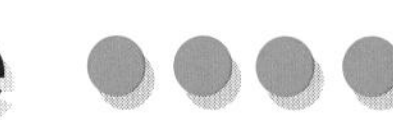

What you need to know

- **Compass points**

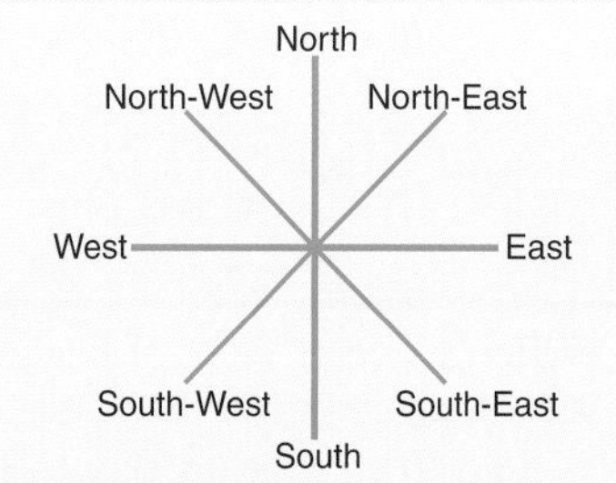

> The angle between North and East is 90°.
> The angle between North and North-East is 45°.

- **Bearings** are used to describe the direction in which you must travel to get from one place to another.
- A bearing is an angle measured from the North line in a clockwise direction.
  A bearing can be any angle from 0° to 360° and is written as a three-figure number.

> To find a bearing:
> measure angle a to find the bearing of Y from X,
> measure angle b to find the bearing of X from Y.

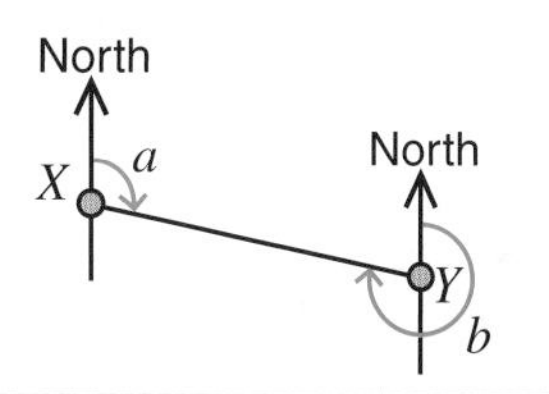

- You should be able to use **scales** and **bearings** to interpret and draw accurate diagrams.

> There are two ways to describe a scale.
> 1. A scale of 1 cm to 10 km means that a distance of 1 cm on the map represents an actual distance of 10 km.
> 2. A scale of 1 : 10 000 means that all distances measured on the map have to be multiplied by 10 000 to find the real distance.

Eg 1 The diagram shows the plan of a stage in a car rally.
The plan has been drawn to a scale of 1 : 50 000.

(a) What is the bearing of Q from P?
(b) What is the bearing of P from R?
(c) What is the actual distance from P to R in metres?

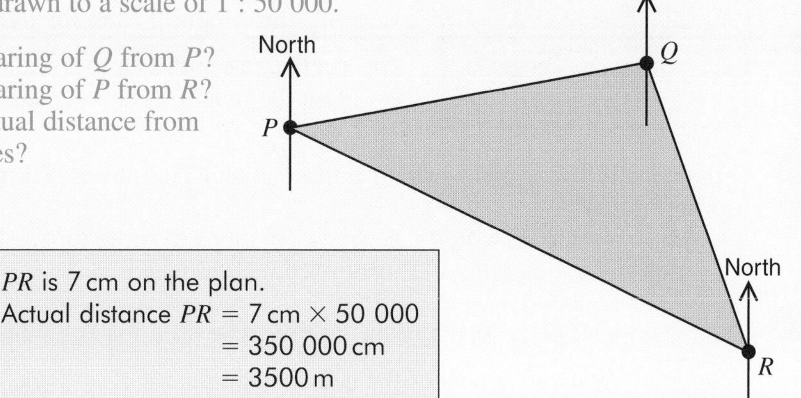

(a) 080°
(b) 295°
(c) 3500 m

> PR is 7 cm on the plan.
> Actual distance PR = 7 cm × 50 000
> = 350 000 cm
> = 3500 m

Exercise 26

1 Jon is facing North-West.
He turns through 180°.
In which direction is he now facing?

2 Here is a map of Jamaica.

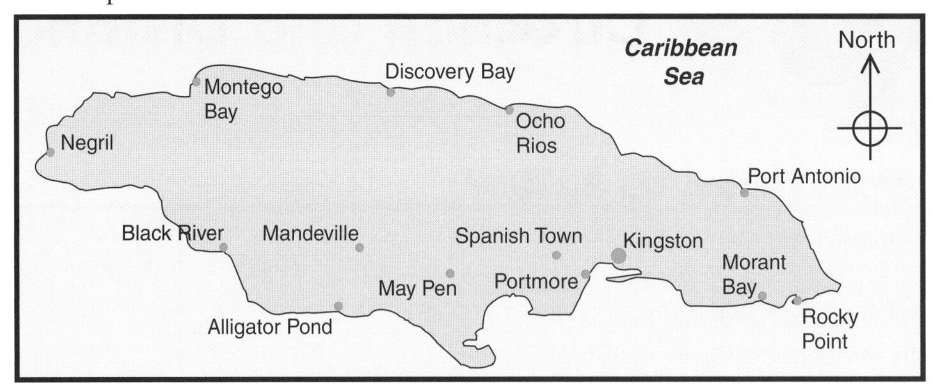

(a) Which place is West of Discovery Bay?
(b) Which place is North-East of Kingston?
(c) You go from Morant Bay to Rocky Point. What compass direction is this? OCR

3 A bridge is 2600 m in length.
A plan of the bridge has been drawn to a scale of 1 cm to 100 m.
What is the length of the bridge on the plan?

4 This map shows three places in Northumberland.

Lanehead (L) Charlton (C) Midhopelaw Pike (M)

Scale: 1 cm to 1 km

(a) What is the real distance from Lanehead to Charlton?
(b) What is the bearing of Midhopelaw Pike from Charlton?

Copy the map.

(c) Bellingham (B) is 6.7 km from C on a bearing of 117°.
Mark the position of B on your map.

M N L C

OCR

5

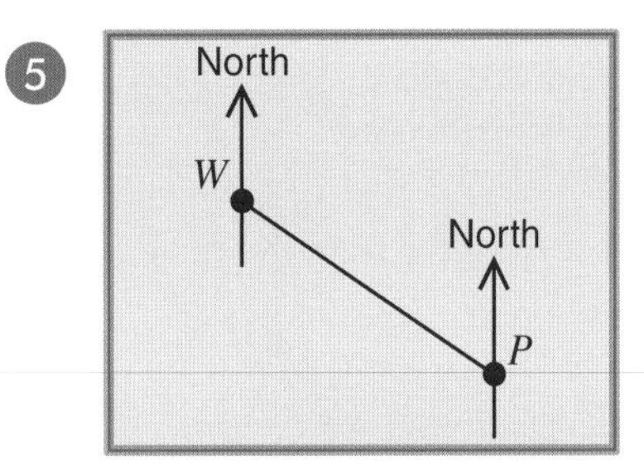

The map shows the positions of a windmill, W, and a pylon, P.

(a) What is the bearing of
(i) the pylon from the windmill,
(ii) the windmill from the pylon?

The map has been drawn to a scale of 2 cm to 5 km.

(b) Use the map to find the distance WP in kilometres.

6 Crossmead is 40 km due North of Fulbridge and Thornby is 30 km from Fulbridge on a bearing of 026°.
(a) Make a scale drawing to show the positions of these towns. Use a scale of 1 cm to 5 km.
(b) Measure the bearing of Thornby from Crossmead. OCR

7 The diagram shows a sketch of the course to be used for a running event.

(a) Draw an accurate plan of the course, using a scale of 1 cm to represent 100 m.

(b) Use your plan to find
(i) the bearing of X from Y,
(ii) the distance XY in metres.

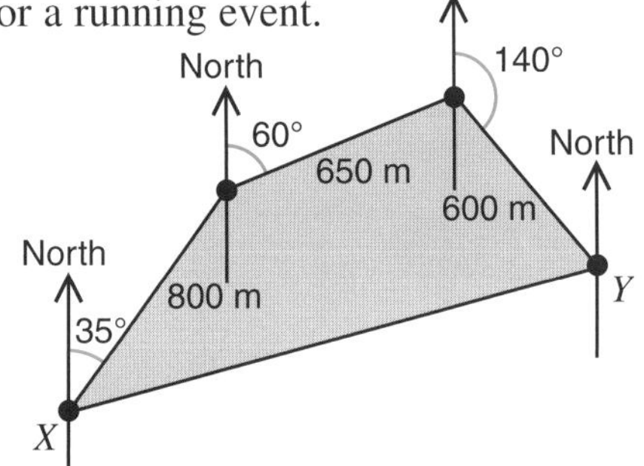

SECTION 27

Circles

What you need to know

- A **circle** is the shape drawn by keeping a pencil the same distance from a fixed point on a piece of paper.
- Words associated with circles:

Circumference – perimeter of a circle.
Radius – distance from the centre of the circle to any point on the circumference. The plural of radius is **radii**.
Diameter – distance right across the circle, passing through the centre point.
Chord – a line joining two points on the circumference.
Tangent – a line which touches the circumference of a circle at one point only. A tangent is perpendicular to the radius at the point of contact.
Arc – part of the circumference of a circle.
Segment – a chord divides a circle into two segments.
Sector – two radii divide a circle into two sectors.

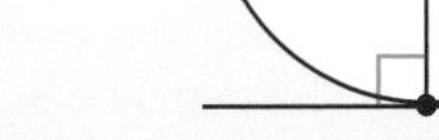

- The **circumference** of a circle is given by: $C = \pi \times d$ or $C = 2 \times \pi \times r$
- The **area** of a circle is given by: $A = \pi \times r^2$
- You should be able to solve problems which involve finding the circumference or the area of a circle.
- Take π to be 3.14 or use the π key on your calculator.

Eg 1 Calculate the circumference of a circle with diameter 18 cm.
Give your answer to 1 d.p.
$C = \pi \times d$
$C = \pi \times 18$
$C = 56.548...$ $\quad C = 56.5$ cm, correct to 1 d.p.

Eg 2 Find the area of a circle with radius 6 cm.
Give your answer to 3 sig. figs.
$A = \pi \times r^2$
$A = \pi \times 6 \times 6$
$A = 113.097...$ $\quad A = 113\text{ cm}^2$, correct to 3 sig. figs.

Eg 3 A circle has a circumference of 25.2 cm.
Find the diameter of the circle.
$C = \pi d$ so, $d = \frac{C}{\pi}$
$d = \frac{25.2}{\pi}$
$d = 8.021...$ $\quad d = 8.0$ cm, correct to 1 d.p.

Eg 4 A circle has an area of 154 cm^2.
Find the radius of the circle.
$A = \pi r^2$ so, $r^2 = \frac{A}{\pi}$
$r^2 = \frac{154}{\pi} = 49.019...$
$r = \sqrt{49.019...} = 7.001...$ $\quad r = 7$ cm, to the nearest cm.

Exercise 27

Do not use a calculator for questions 1 and 2.

1 This is a full-size drawing of a hole cut in a metal sheet.

(a) (i) Find the radius of the hole.

(ii) Calculate the area of the hole.
Write down your calculations.

(b) Calculate the circumference of the hole. OCR

2 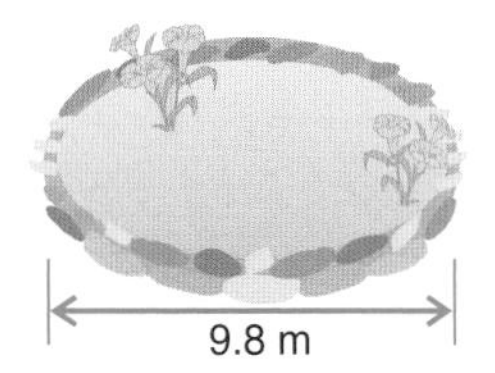

A circular pond has a diameter of 9.8 metres.

(a) Estimate the circumference of the pond.

(b) Estimate the area of the pond.

3 A circle has a diameter of 5 cm.

(a) Calculate the circumference of the circle.

(b) Calculate the area of the circle.

Give your answers correct to one decimal place.

4 Discs of card are used in the packaging of frozen pizzas.
Each disc fits the base of the pizza exactly.
Calculate the area of a disc used to pack a pizza.
Give your answer in terms of π.

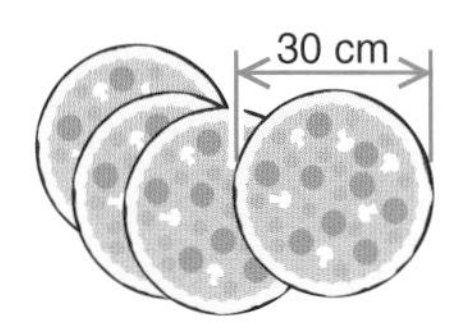

5 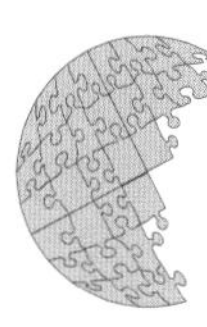

Tranter has completed three-fifths of a circular jigsaw puzzle.
The puzzle has a radius of 20 cm.
What area of the puzzle is complete?

6 Mr Kray's lawn is 25 m in length.
He rolls it with a garden roller.
The garden roller has a diameter of 0.4 m.
Work out the number of times the roller rotates when rolling the length of the lawn once.

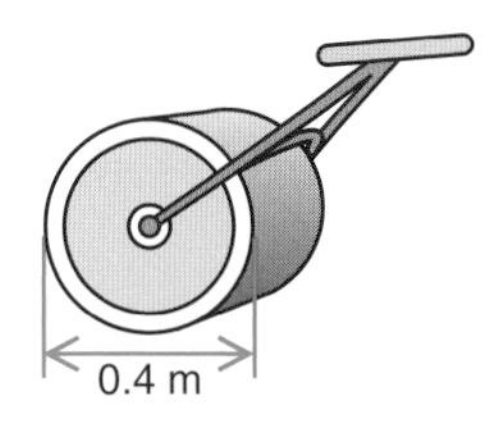

7 

The top of a table is a circle with a radius of 55 cm.

(a) Calculate the circumference of the table top.

On the table are 6 place mats.
Each place mat is a circle with a diameter of 18 cm.

(b) What area of the table top is **not** covered by place mats?

8 The logo for a football team is a white circle with a blue design inside.
The team paint their logo on the pitch.
The circle has a radius of 5 m.
The blue area covers 35% of the circle.
Find the area painted blue.

OCR

9 This design is made with three semi-circles, each of diameter 8 cm.
Find the perimeter of the design.

OCR

10 A circle has a circumference of 100 cm.
Calculate the area of the circle. Give your answer correct to three significant figures.

11 Alfie says, ***"A semi-circle with a radius of 10 cm has a larger area than a whole circle with half the radius."***

Is he correct? You **must** show working to justify your answer.

SECTION 28 Areas and Volumes

What you need to know

- Shapes formed by joining different shapes together are called **compound shapes**.
  To find the area of a compound shape we must first divide the shape up into rectangles, triangles, circles, etc., and find the area of each part.
  Add the answers to find the total area.

  Eg 1 Find the total area of this shape.

  Area $A = 5 \times 4 = 20\,\text{cm}^2$
  Area $B = 6 \times 3 = 18\,\text{cm}^2$
  Total area $= 20 + 18 = 38\,\text{cm}^2$

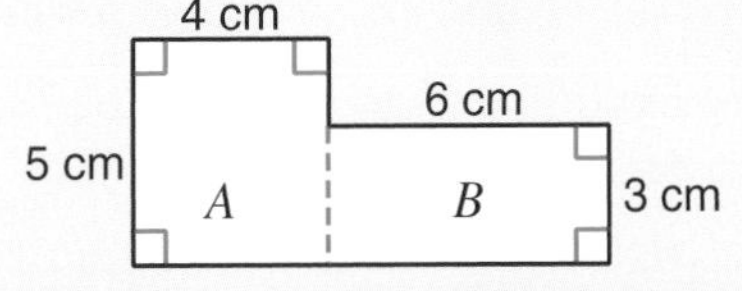

- **Faces**, **vertices** (corners) and **edges**.

  Eg 2 A cube has 6 faces, 8 vertices and 12 edges.

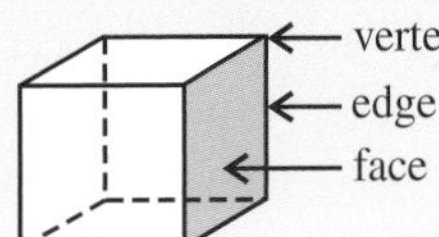

- A **net** can be used to make a solid shape.

  Eg 3 Draw a net of a cube.

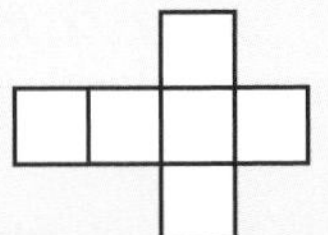

- **Isometric paper** is used to make 2-D drawings of 3-D shapes.

  Eg 4 Draw a cube of edge 2 cm on isometric paper.

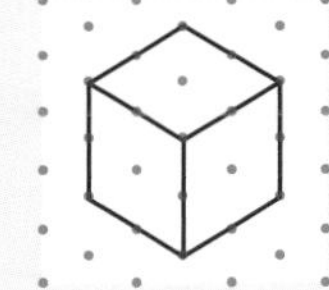

- **Plans and Elevations**
  The view of a 3-D shape looking from above is called a **plan**.
  The view of a 3-D shape from the front or sides is called an **elevation**.

  Eg 5 Draw diagrams to show the plan and elevation from **X**, for this 3-dimensional shape.

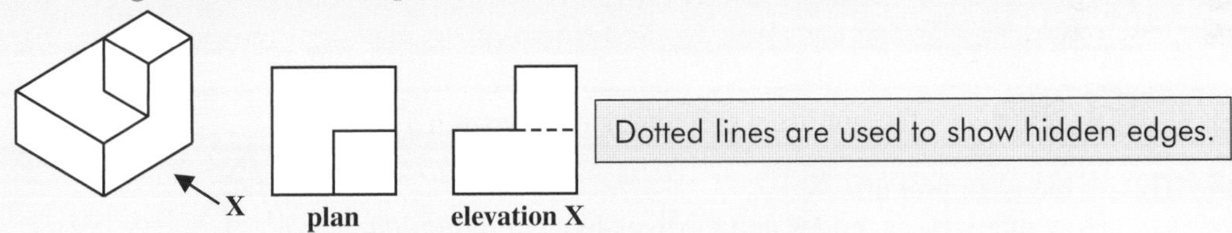

- **Volume** is the amount of space occupied by a 3-dimensional shape.
- The formula for the volume of a **cuboid** is:
  Volume = length $\times$ breadth $\times$ height
  $V = l \times b \times h$

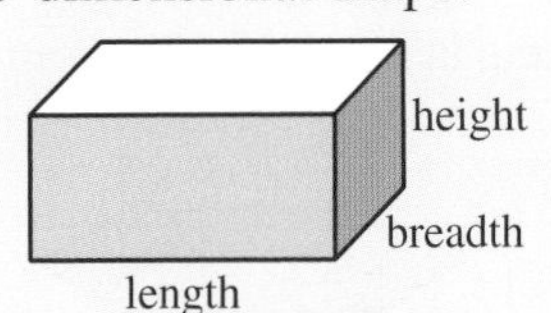

- Volume of a **cube** is: $V = l^3$
- To find the **surface area** of a cuboid, find the areas of the 6 rectangular faces and add the answers together.

  Eg 6 Find the volume and surface area of a cuboid measuring 7 cm by 5 cm by 3 cm.

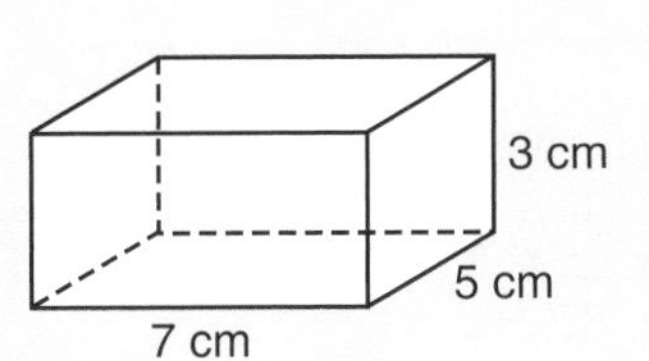

Volume $= l \times b \times h$
$= 7\,\text{cm} \times 5\,\text{cm} \times 3\,\text{cm}$
$= 105\,\text{cm}^3$

Surface area $= (2 \times 7 \times 5) + (2 \times 5 \times 3) + (2 \times 3 \times 7)$
$= 70 + 30 + 42$
$= 142\,\text{cm}^2$

Eg 7 This cuboid has a volume of $75\,\text{cm}^3$.
Calculate the height, h, of the cuboid.

Volume $= lbh$

$75 = 6 \times 5 \times h$

$h = \frac{75}{30}$

$h = 2.5\,\text{cm}$

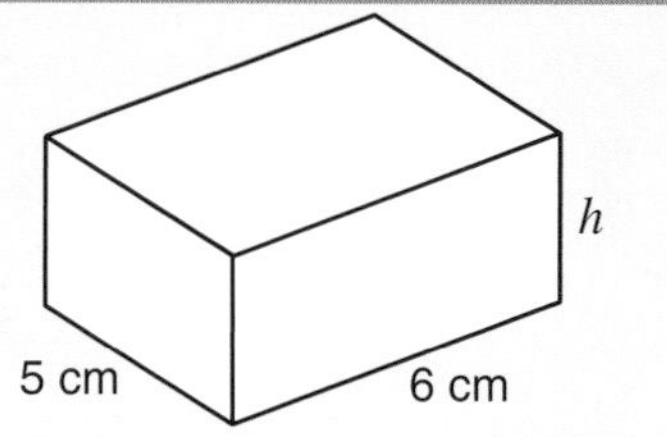

- **Prisms**
  If you make a cut at right angles to the length of a prism you will always get the same cross-section.
- Volume of a prism = area of cross-section × length

Triangular prism
cross-section
length

Eg 8 Calculate the volume of this prism.
The cross-section of this prism is a trapezium.

Area of cross-section $= \frac{1}{2}(5 + 3) \times 2.5$
$= 4 \times 2.5$
$= 10\,\text{cm}^2$

Volume of prism = area of cross-section × length
$= 10 \times 12$
$= 120\,\text{cm}^3$

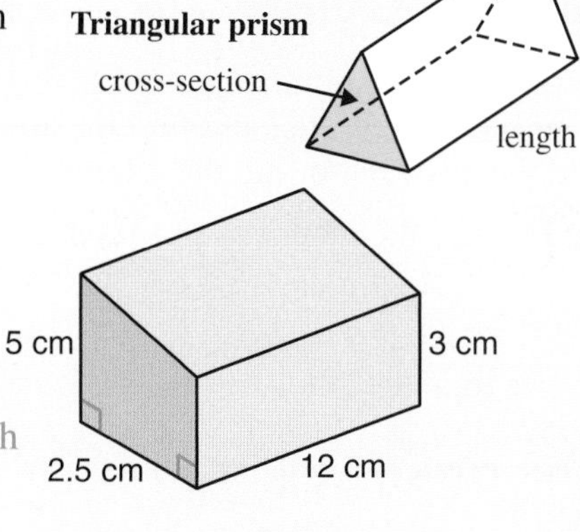

- A **cylinder** is a prism.
  Volume of a cylinder is: Volume $= \pi \times r^2 \times h$
  Surface area of a cylinder is: Surface area $= 2\pi r^2 + 2\pi rh$

r
h

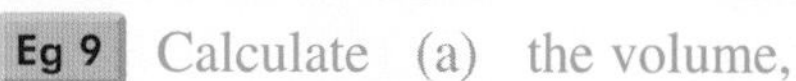

Eg 9 Calculate (a) the volume,
(b) the surface area of this cylinder.

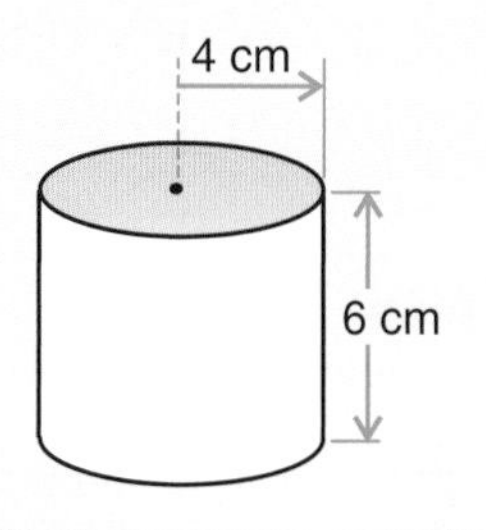

(a) Volume $= \pi r^2 h$
$= \pi \times 4 \times 4 \times 6$
$= 301.592\ldots$
$= 302\,\text{cm}^3$, correct to 3 s.f.

(b) Surface area $= 2\pi r^2 + 2\pi rh$
$= 2 \times \pi \times 4 \times 4 + 2 \times \pi \times 4 \times 6$
$= 100.53\ldots + 150.796\ldots$
$= 251\,\text{cm}^2$, correct to 3 s.f.

Exercise 28

Do not use a calculator for question 1 to 7.

1. This shape is a pyramid.
   (a) How many faces, edges and vertices has the pyramid?
   (b) Which of these nets is a net of the pyramid?

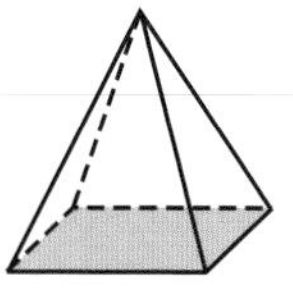

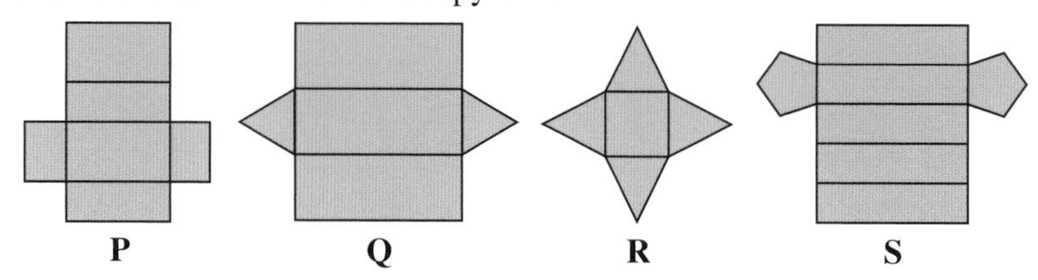

2. The shapes are drawn on a grid with 1 cm squares.
   (a) Work out the area on the 'T' shape.
   (b) Estimate the area of the oval shape.
   (c) Work out the perimeter of the 'T' shape.

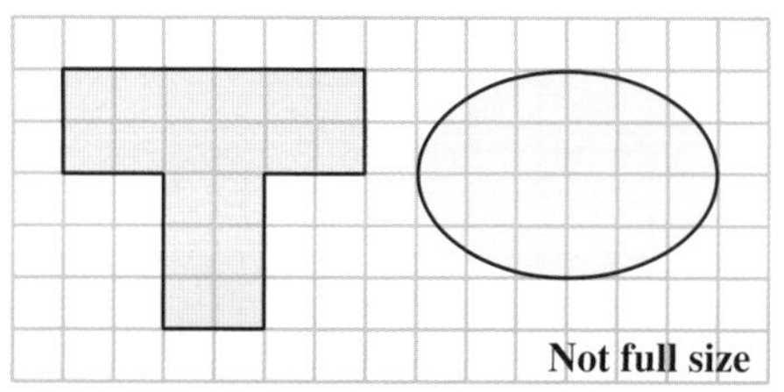

OCR

3 Find the area of this shape.

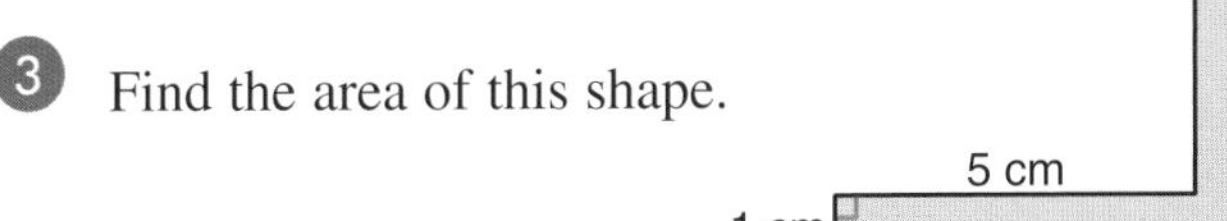

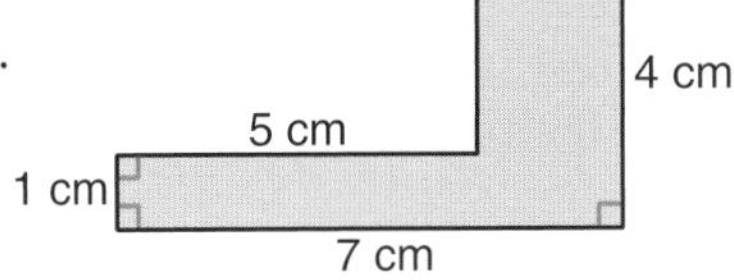

4 This cuboid has been made using cubes of side 1 cm.

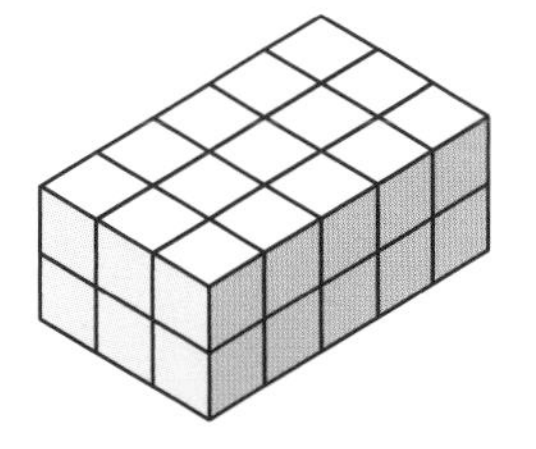

(a) How many cubes are needed to make the cuboid?

(b) (i) Draw a net of the cuboid on 1 cm squared paper.

(ii) Hence, find the surface area of the cuboid.

5 Kanta has lots of these building blocks. There are three types, **A**, **B** and **C**.

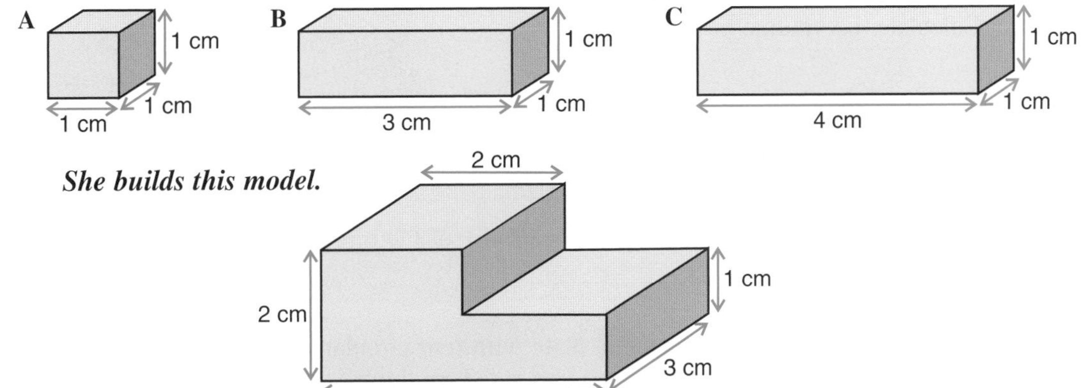

(a) (i) First she builds the model using only type **B** blocks. How many does she need?

(ii) Then she builds it using type **B** and type **C** blocks. How many of each does she need?

(b) (i) What is the volume of the type **B** block?

(ii) What is the volume of the model?

OCR

6 The diagram shows a solid drawn on isometric paper.

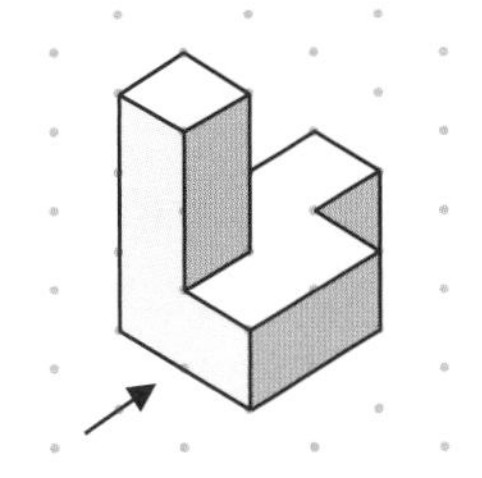

(a) Draw the plan of the solid.

(b) Draw the elevation of the solid from the direction shown by the arrow.

7 Which of these cuboids has the largest volume? Show all your working.

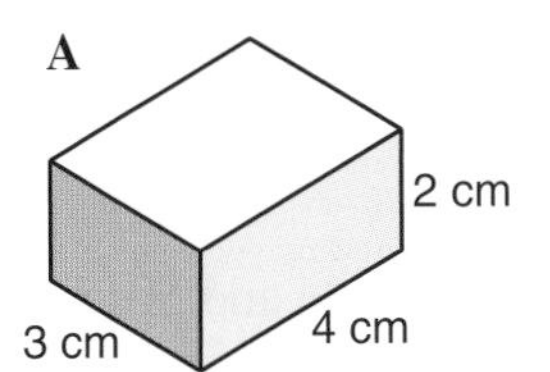

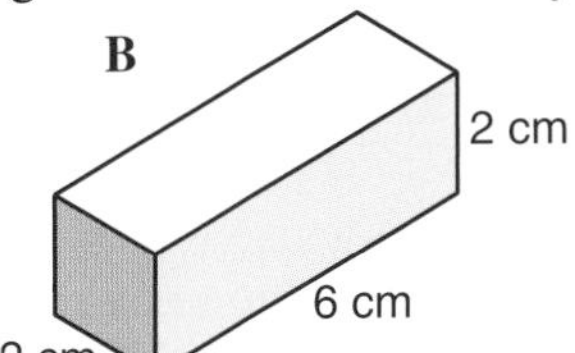

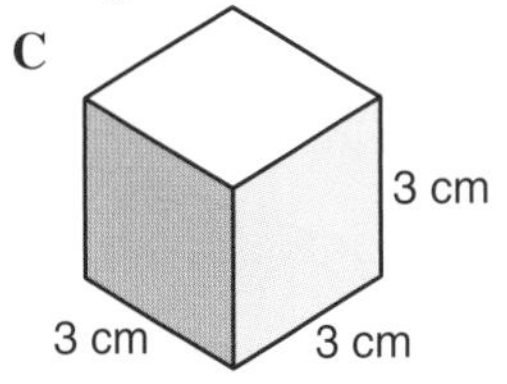

8 A chest freezer is a cuboid. The freezer is 1.5 m wide, 0.6 m deep and 1.2 m high.

(a) Calculate the volume of the freezer.

(b) 72% of the volume of the freezer can be used for food storage. Calculate the volume that can be used for food storage.

OCR

9 (a) The base of a cuboid has length 16 cm and width 12.5 cm. The volume of the cuboid is 1880 cm^3. Find the height of the cuboid.

(b) The volume of another cuboid is 36 cm^3. The length, width and height of the cuboid are all different whole numbers. Give one set of possible values of the length, width and height.

OCR

10 A photo frame is a square of side 20 cm.
It has a circular glass section and the rest is brass.
Work out the area of the brass part of the photo frame.

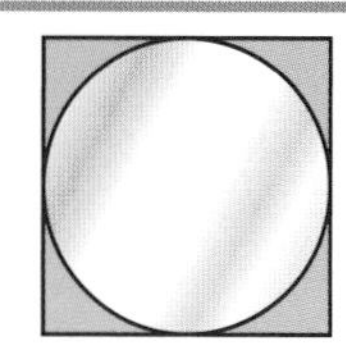

OCR

11

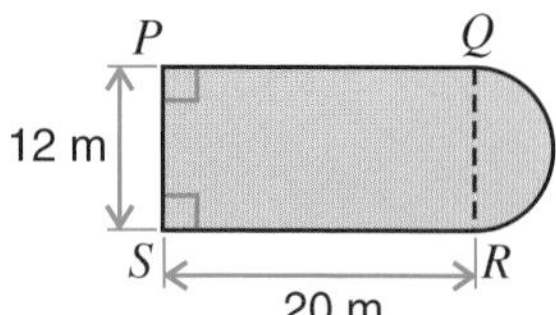

The diagram shows the plan of a swimming pool.
The arc QR is a semi-circle.
$PS = 12\text{ m}$ and $PQ = RS = 20\text{ m}$.
Calculate the area of the surface of the pool.

12 A triangular prism has dimensions, as shown.
(a) Calculate the total surface area of the prism.
(b) Calculate the volume of the prism.

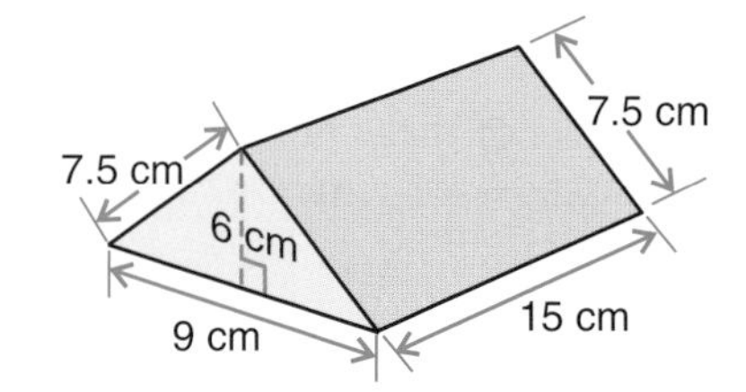

13

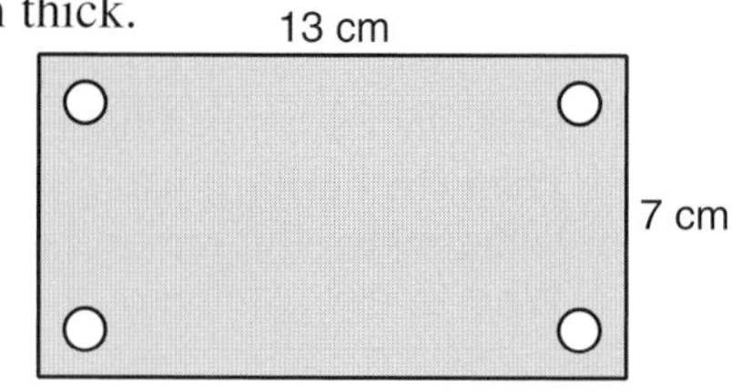

The diagram shows a grit bin.
The bin is a prism.

Calculate the capacity of the bin.

OCR

14 The diagram shows a rectangular metal plate with four circular holes.
The metal plate measures 13 cm by 7 cm and is 0.3 cm thick.
The radius of each circle is 0.4 cm.

Calculate the volume of the metal.

15

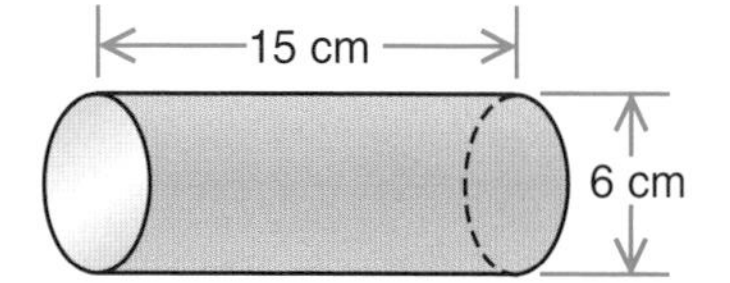

The diagram shows a cylinder.

Calculate the volume of the cylinder.

16 (a) The diagram shows the cross-section of a water trough.
It is in the shape of a semi-circle with radius 12 cm.
The trough is 84 m long.
What volume of water can the trough hold?
(b) Convert your answer in part (a) to litres.

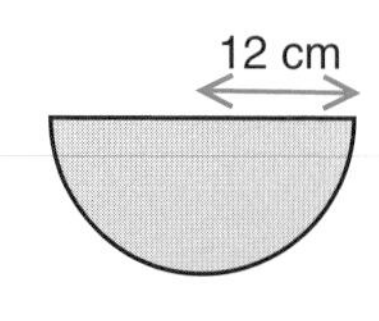

OCR

17

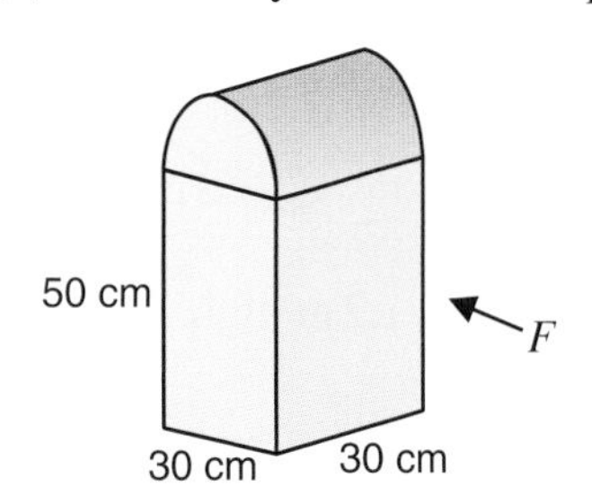

A kitchen waste bin is a prism.
The cross-section is a rectangle 30 cm wide and 50 cm high, topped by a semi-circle of radius 15 cm.
The bin has a square base.
(a) Draw the plan and the front elevation viewed from F.
Use a scale of 1 cm to 10 cm.
(b) Calculate the total volume of the waste bin.

OCR

18 A cylindrical water tank has radius 40 cm and height 90 cm.
(a) Calculate the total surface area of the tank.

A full tank of water is used to fill a paddling pool.
(b) The paddling pool is a square based prism, as shown.
Calculate the depth of water in the pool.

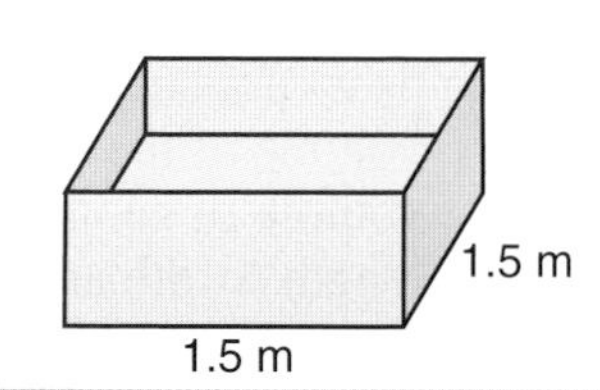

SECTION 29

Loci and Constructions

What you need to know

- The path of a point which moves according to a rule is called a **locus**.
- The word **loci** is used when we talk about more than one locus.
- You should be able to draw the locus of a point which moves according to a given rule.

Eg 1 A ball is rolled along this zig-zag. Draw the locus of *P*, the centre of the ball, as it is rolled along.

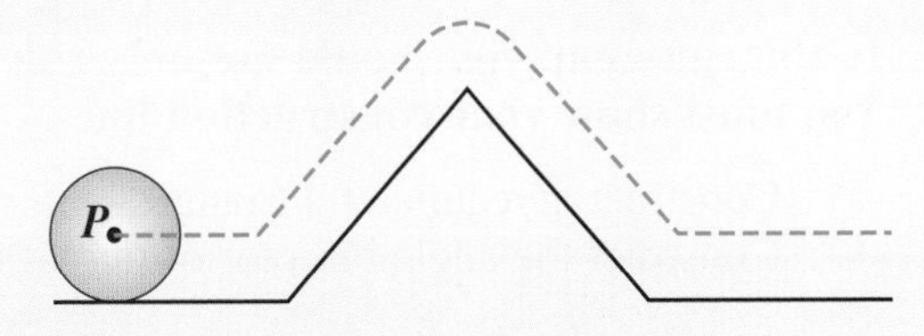

- Using a ruler and compasses you should be able to carry out the **constructions** below.

1 The perpendicular bisector of a line.

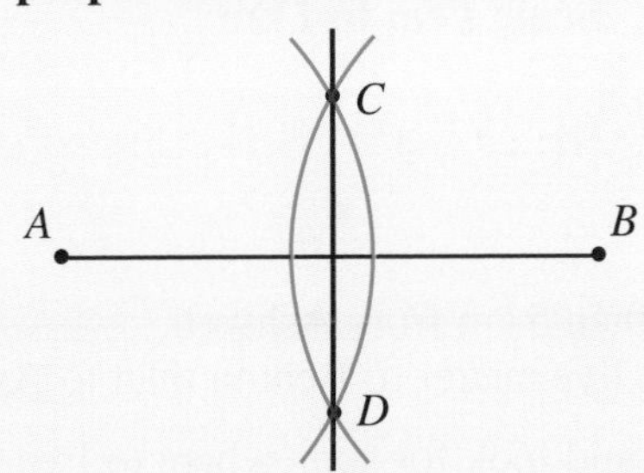

Points on the line *CD* are **equidistant** from the points *A* and *B*.

2 The bisector of an angle.

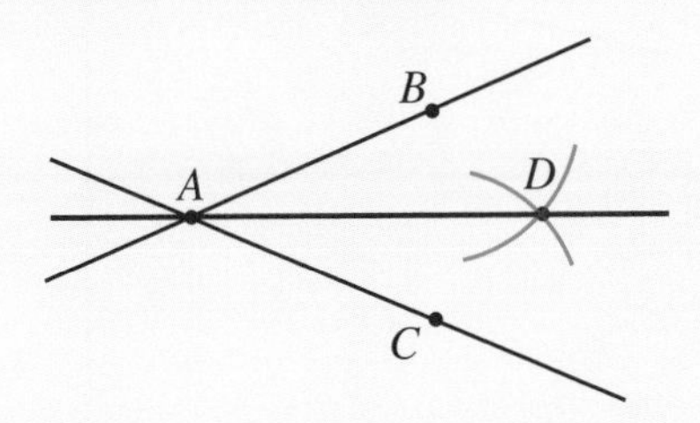

Points on the line *AD* are **equidistant** from the lines *AB* and *AC*.

3 The perpendicular from a point to a line.

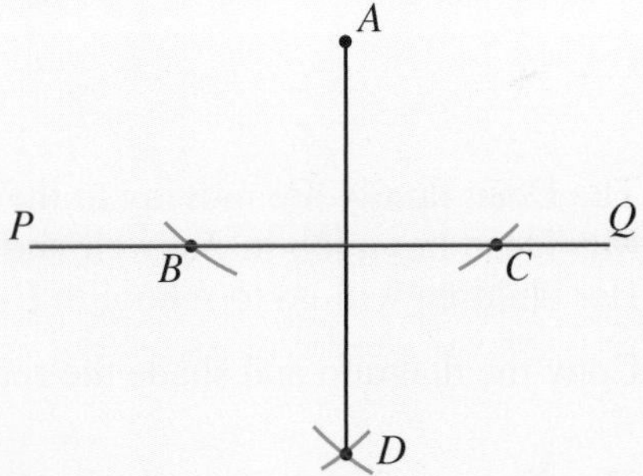

4 The perpendicular from a point on a line.

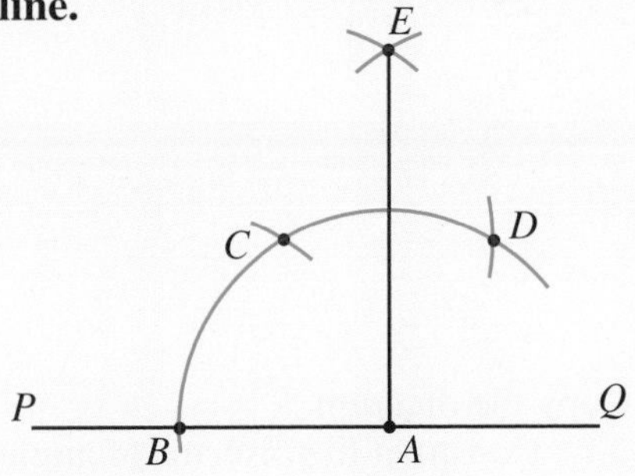

- You should be able to solve loci problems which involve using these constructions.

Eg 2 *P* is a point inside triangle *ABC* such that:
(i) *P* is equidistant from points *A* and *B*,
(ii) *P* is equidistant from lines *AB* and *BC*.
Find the position of *P*.

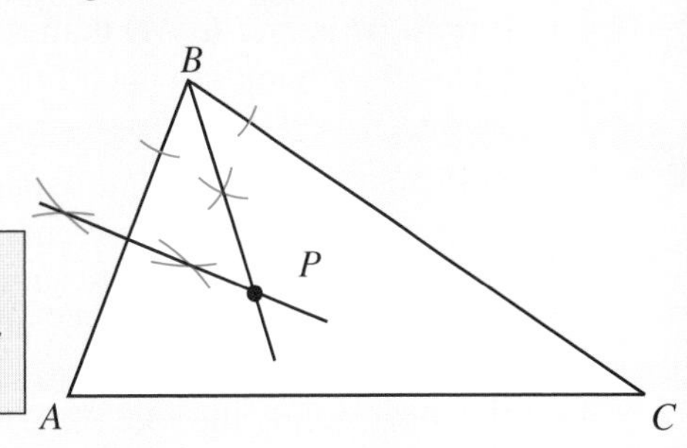

To find point *P*:
(i) construct the perpendicular bisector of line *AB*,
(ii) construct the bisector of angle *ABC*.

P is at the point where these lines intersect.

Exercise 29

1 The ball is rolled along the zig-zag.
Copy the diagram and draw the locus of the centre of the ball as it is rolled from X to Y.

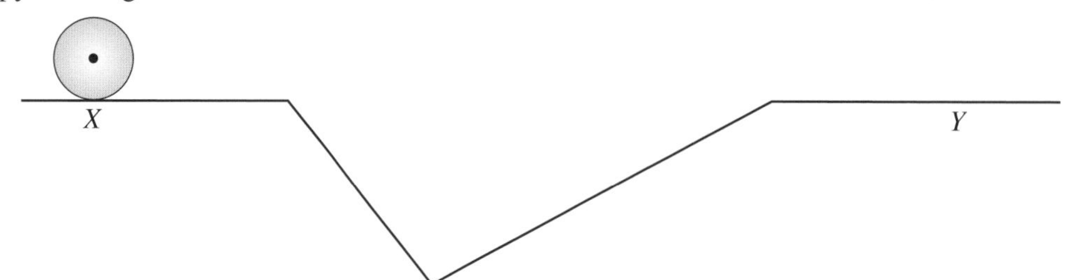

2 **In this question, you should use only ruler and compasses.**
You must show your construction lines.

(a) Construct an equilateral triangle with sides of length 7 cm.
(b) Construct the angle bisector of one of the angles of your triangle. OCR

3 This scale drawing shows two villages, Ashwell (A) and Benton (B). They are 8 km apart.

A

Scale 1 cm to 1 km

B

The Dean family are moving to the area.
Mrs Dean has a job in Ashwell and wants to live less than 5 km from Ashwell.
Her children will go to school in Benton, so they must live nearer to Benton than to Ashwell.

Copy the diagram and shade the region where they should look for somewhere to live. OCR

4 The diagram shows the scale drawing of a garden, $ABCD$.

The scale is **1 cm to 2 m**.

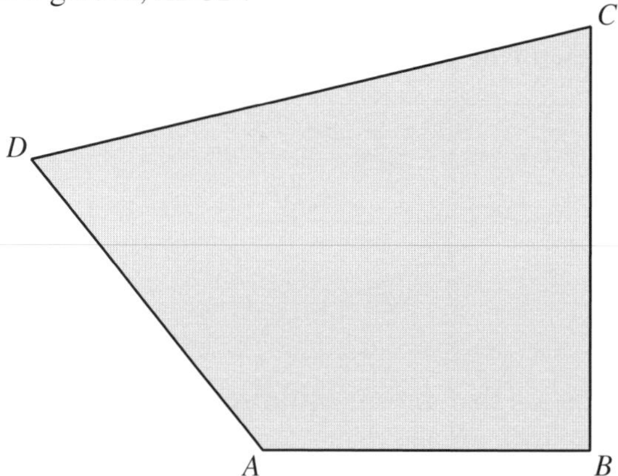

Copy the diagram.
(a) Construct the bisector of angle A.
Show all your construction lines.
(b) A tree is to be planted in the garden.
It must be nearer to AD than AB and at least 8 m from C.
Shade the region in which the tree could be planted. OCR

5 (a) Construct a kite $PQRS$ in which $PQ = PS = 7$ cm, $QR = RS = 5$ cm and the diagonal $QS = 6$ cm.
X is a point inside the kite such that:
(i) X is equidistant from P and Q,
(ii) X is equidistant from sides PQ and PS.
(b) By constructing the loci for (i) and (ii) find the position of X.
(c) Measure the distance PX.

SECTION 30 Transformations

What you need to know

- The movement of a shape from one position to another is called a **transformation**.
- **Single transformations** can be described in terms of a reflection, a rotation or a translation.
- **Reflection**: The image of the shape is the same distance from the mirror line as the original.

Eg 1 Reflect shape P in the line AB.

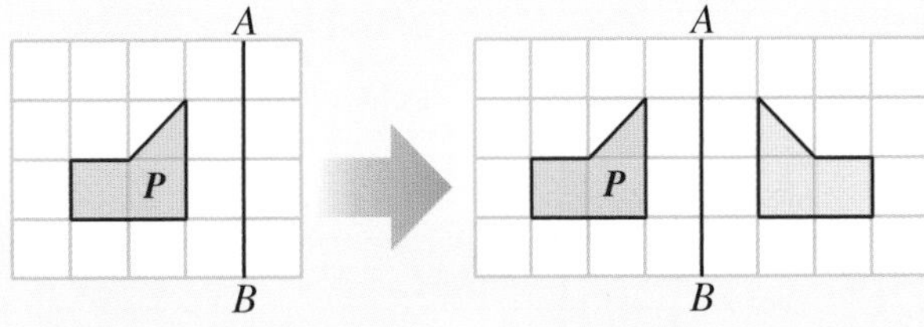

- **Rotation**: All points are turned through the same angle about the same point, called a centre of rotation.

Eg 2 Rotate shape P 90° clockwise about the origin.

Clockwise means:

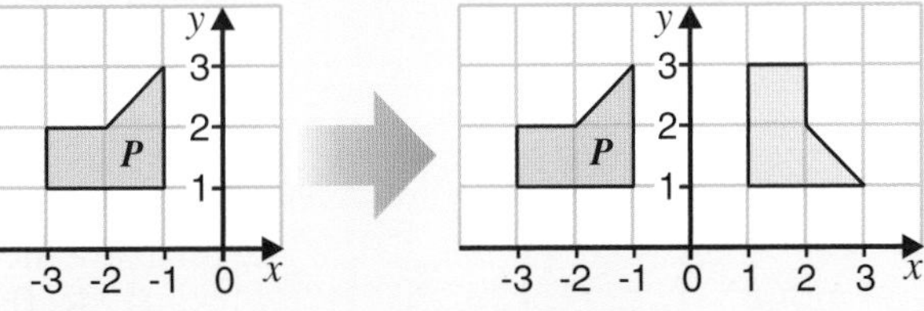

- **Translation**: All points are moved the same distance in the same direction without turning.

Eg 3 Translate shape P with vector $\begin{pmatrix} 3 \\ 1 \end{pmatrix}$.

$\begin{pmatrix} 3 \\ 1 \end{pmatrix}$ means 3 units right and 1 unit up.

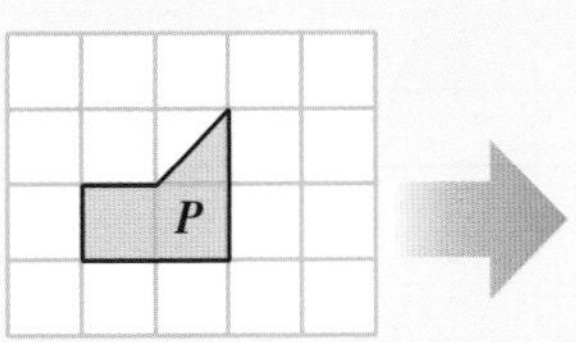

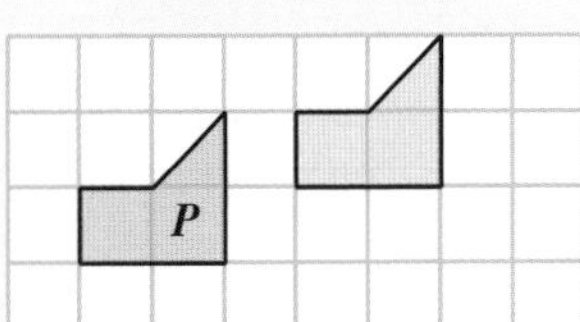

- You should be able to fully describe transformations.

Transformation	Image same shape and size?	Details needed to describe the transformation
Reflection	Yes	Mirror line, sometimes given as an equation.
Rotation	Yes	Centre of rotation, amount of turn, direction of turn.
Translation	Yes	Horizontal movement and vertical movement. Vector: top number = horizontal movement, bottom number = vertical movement.

Eg 4 Describe the single transformation which maps

(a) A onto B,
(b) C onto A,
(c) A onto D.

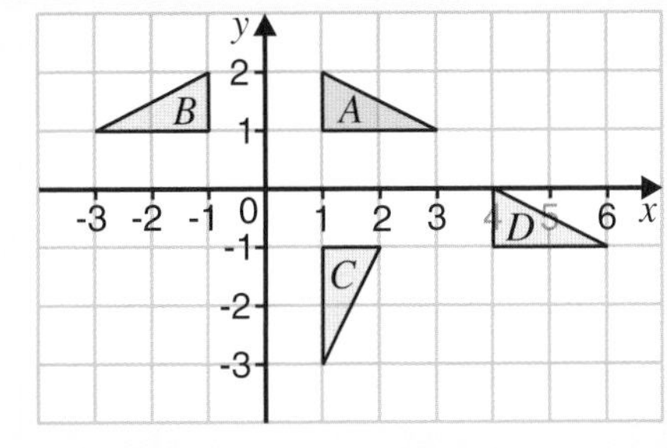

(a) **Reflection** in the y axis.
(b) **Rotation** of 90° anticlockwise about the origin.
(c) **Translation** 3 units to the right and 2 units down.

Exercise 30

1 Copy each diagram and draw the transformation given.

(a) Reflect the shape in the x axis.

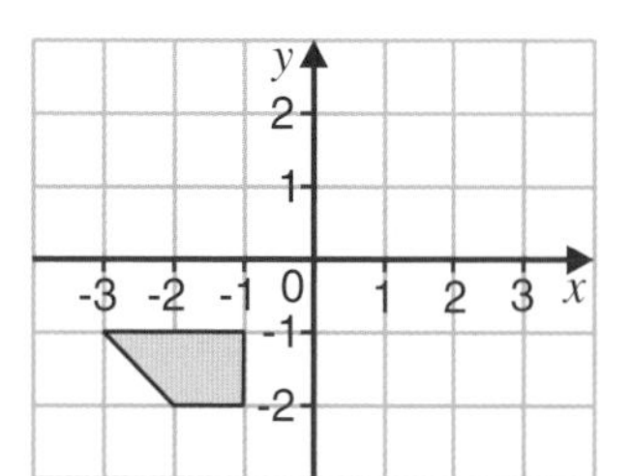

(b) Translate the shape 2 units left and 3 units up.

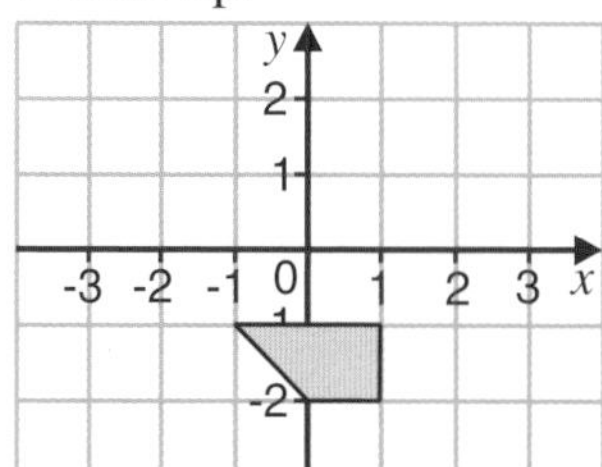

(c) Rotate the shape 90° clockwise about the origin.

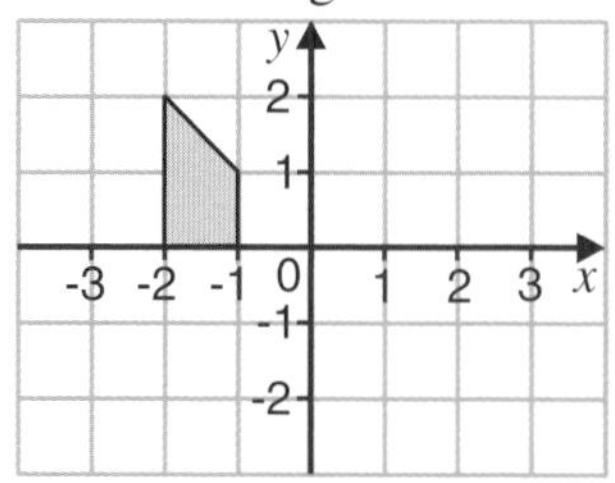

2 In each diagram A is mapped onto B by a single transformation. Describe each transformation.

(a)

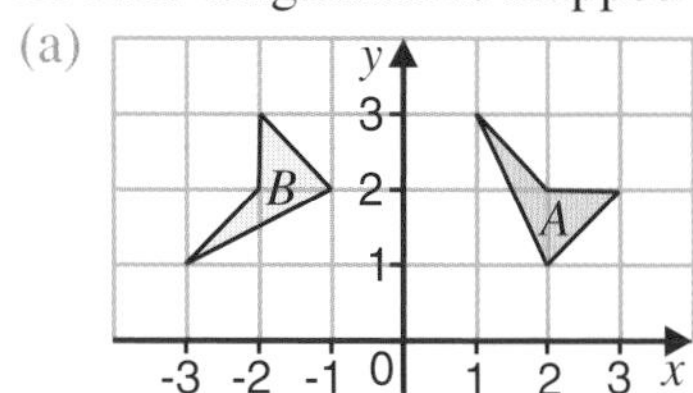

(b)

(c)

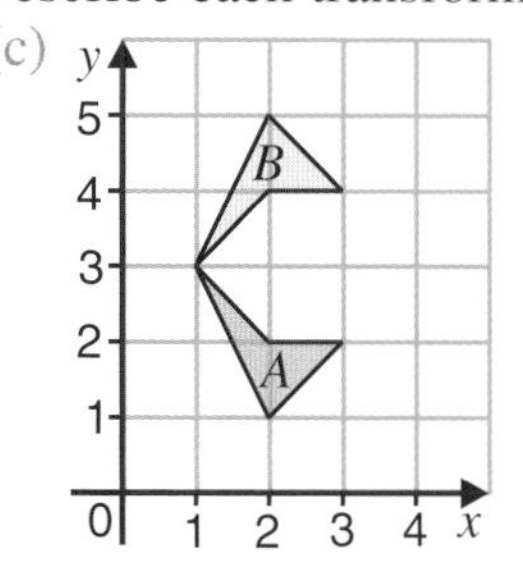

3 (a)

Copy the diagram.

(i) Reflect triangle A in the y axis. Label the reflection B.

(ii) Reflect triangle A in the dotted line. Label the reflection C.

(b) Copy the diagram. Rotate triangle T through 90° anticlockwise about the origin.

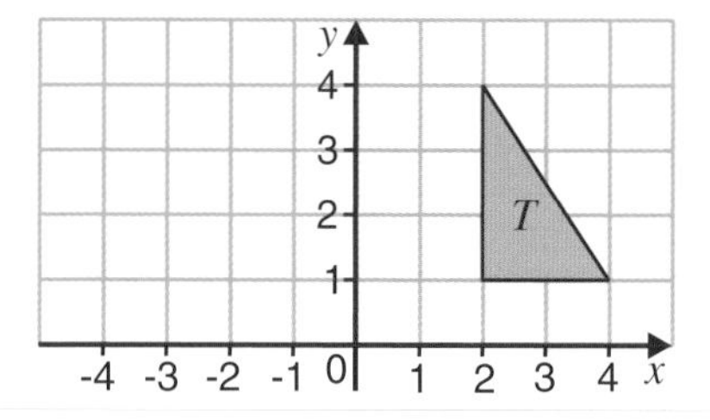

(c) Copy the diagram. Translate triangle P by $\begin{pmatrix} 3 \\ -4 \end{pmatrix}$.

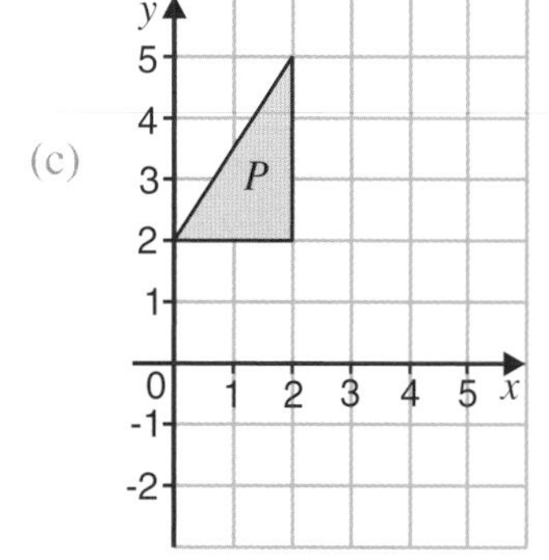

OCR

4 The diagram shows the positions of kites P, Q and R.

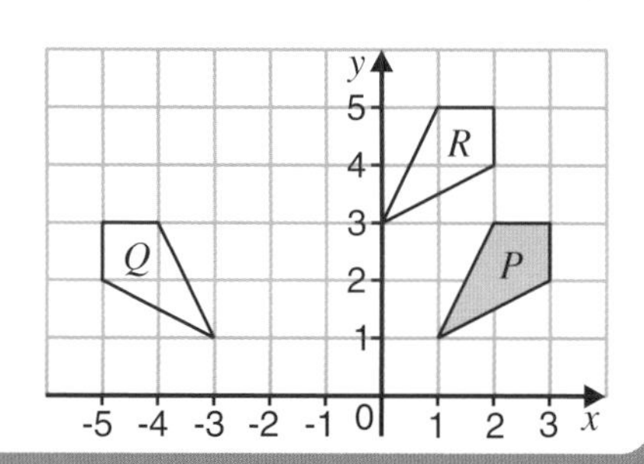

(a) P is mapped onto Q by a reflection. What is the equation of the line of reflection?

(b) P is mapped onto R by a translation. Describe the translation.

(c) P is mapped onto T by a rotation through 90° clockwise about (0, 3). On squared paper, copy P and draw the position of T.

5

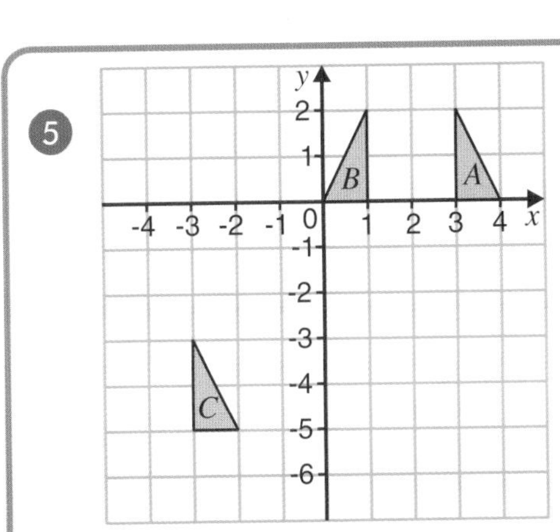

(a) Describe the transformation which maps
  (i) shape A onto shape B,
  (ii) shape A onto shape C.

(b) Copy shape A onto squared paper.
Rotate shape A through 90° clockwise about the origin.
Label the image R.

OCR

6

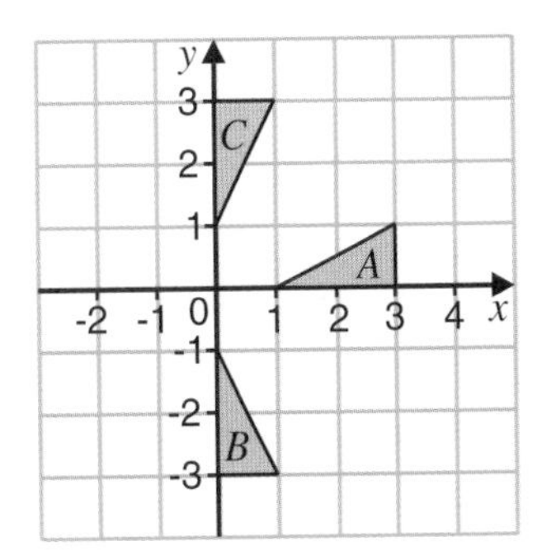

(a) Describe fully the single transformation which maps
  (i) triangle A onto triangle B,
  (ii) triangle A onto triangle C.

(b) Copy shape C onto squared paper.
Translate triangle C by the vector $\begin{pmatrix} -3 \\ 2 \end{pmatrix}$.
Label the answer D.

OCR

7

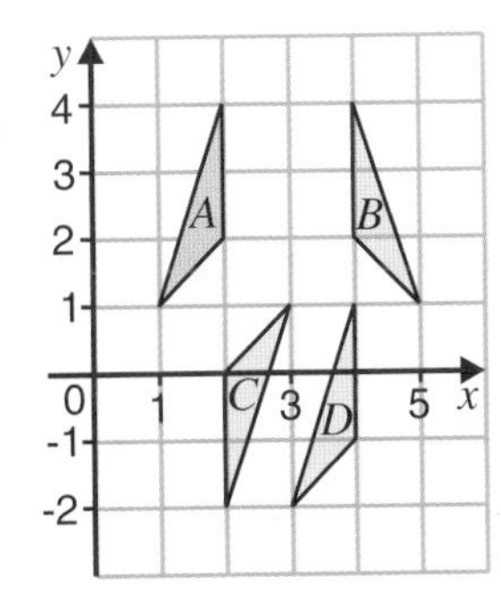

Describe the single transformation which maps
(a) A onto B,
(b) A onto C,
(c) A onto D.

8

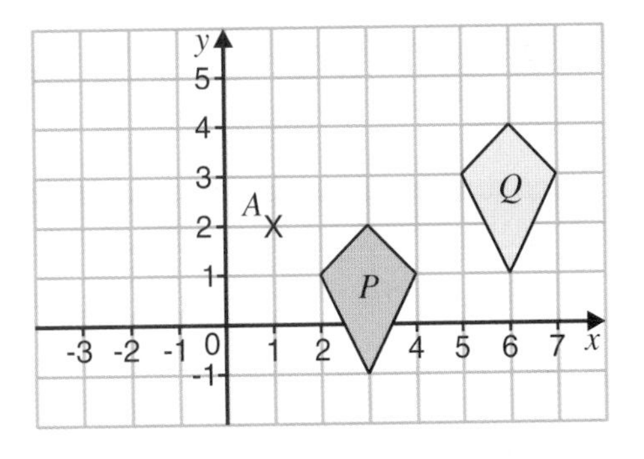

(a) Describe fully the single transformation which maps shape P onto shape Q.

Copy shape P onto squared paper.
(b) Rotate shape P through 90° anticlockwise about the point $A\,(1, 2)$.

9

(a)

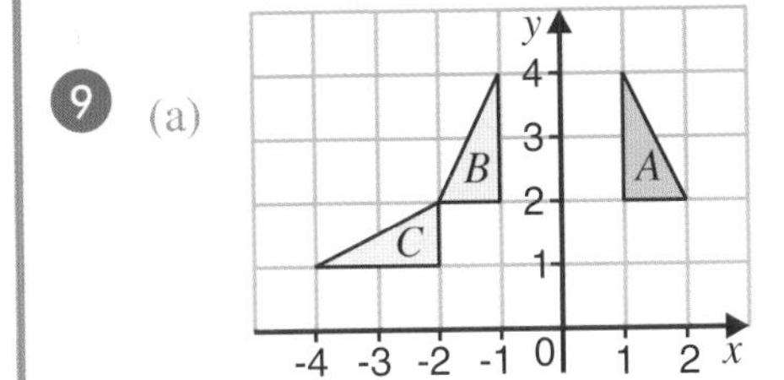

Describe fully the single transformation that maps
  (i) A onto B,
  (ii) A onto C.

(b) Q is reflected in $x = 4$ and then in $y = 1$.
Describe fully the single transformation that is equivalent to these two transformations.

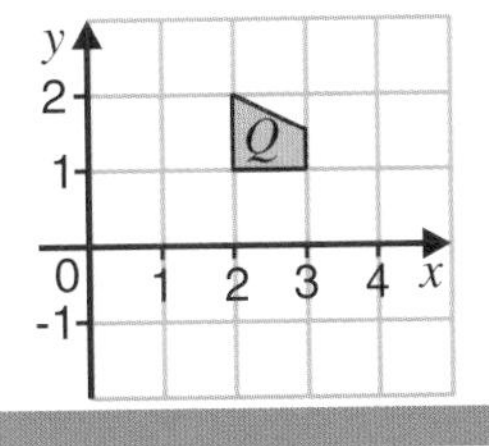

OCR

Enlargements and Similar Figures

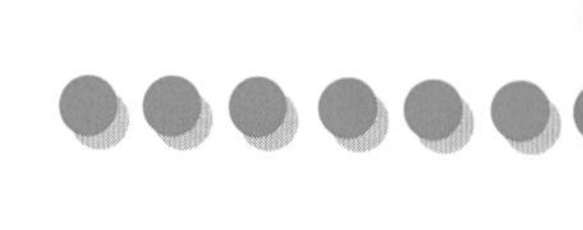

What you need to know

- When a shape is **enlarged**: all **lengths** are multiplied by a **scale factor**,
  angles remain unchanged.
  New length = scale factor × original length.

  The size of the original shape is:
  increased by using a scale factor greater than 1,
  reduced by using a scale factor which is a fraction, i.e. between 0 and 1.

- You should be able to draw an enlargement.

  Eg 1 Draw an enlargement of shape P, with scale factor 2, centre O.

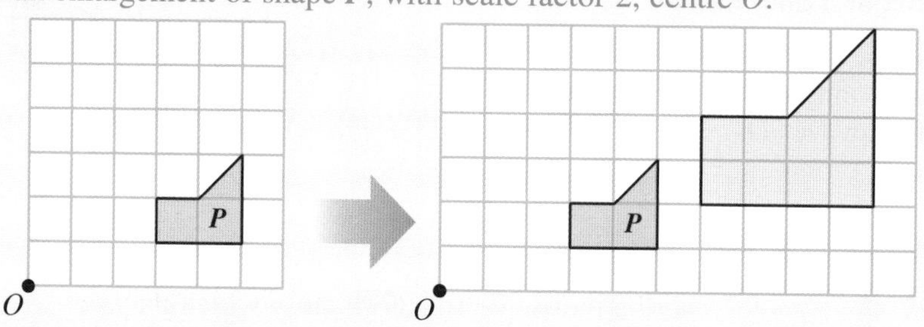

- You should be able to describe an enlargement.

  Eg 2 Describe fully the enlargement which maps P onto Q.

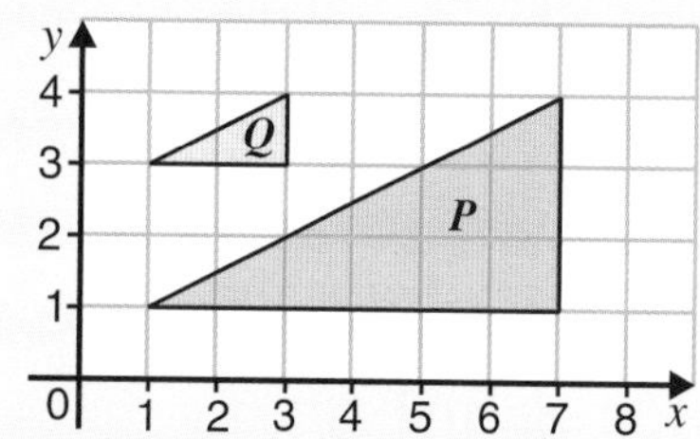

Scale factor $= \dfrac{\text{new length}}{\text{original length}}$

The centre of enlargement is the point where lines drawn through corresponding vertices of shapes P and Q cross.

Enlargement, scale factor $\frac{1}{3}$, centre (1, 4).

- When two figures are **similar**:
  their **shapes** are the same, their **angles** are the same,
  corresponding **lengths** are in the same ratio, this ratio is the **scale factor** of the enlargement.

  Scale factor $= \dfrac{\text{new length}}{\text{original length}}$ New length = scale factor × original length

- All circles are similar to each other.
- All squares are similar to each other.
- You should be able to find corresponding lengths in similar shapes.

  Eg 3 These two shapes are similar.
  (a) Find the lengths of the sides marked x and y.
  (b) Find angle PQR.

  AB and PQ are corresponding sides.
  Scale factor $= \dfrac{PQ}{AB} = \dfrac{5}{3}$

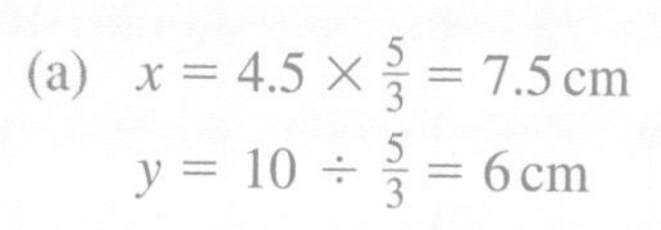

(a) $x = 4.5 \times \frac{5}{3} = 7.5\,\text{cm}$

$y = 10 \div \frac{5}{3} = 6\,\text{cm}$

(b) Angles stay the same.
$\angle PQR = 100°$

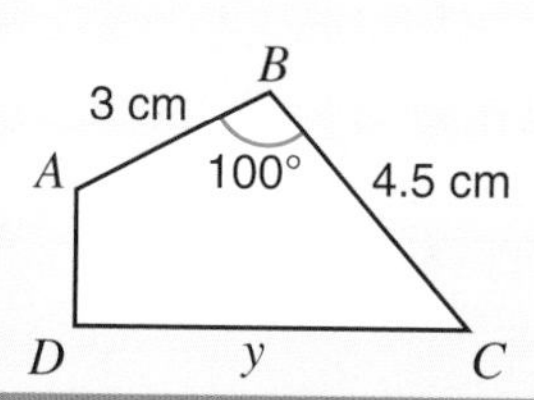

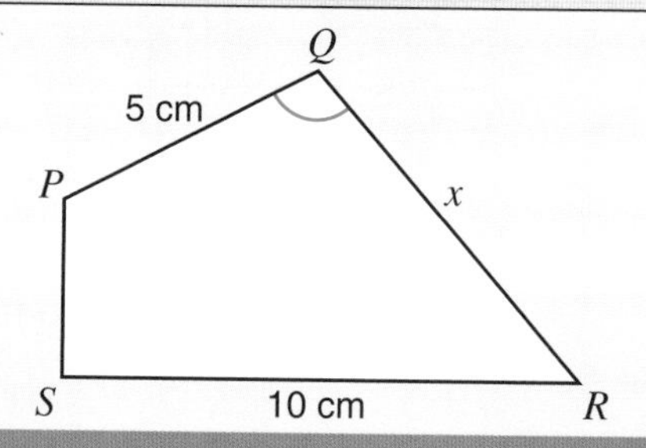

Exercise 31

1 Copy the shape onto squared paper.
Enlarge this shape using a scale factor of 3.
Use C as the centre of enlargement.

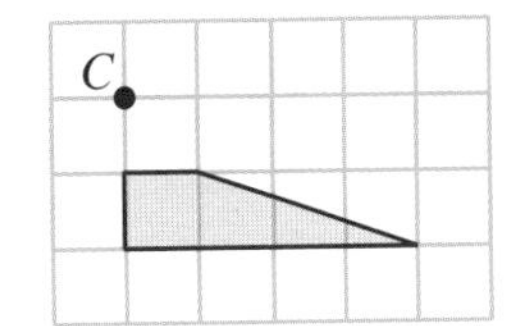

OCR

2 Enlarge shape **A** with centre $(-1, 0)$ and scale factor 2.

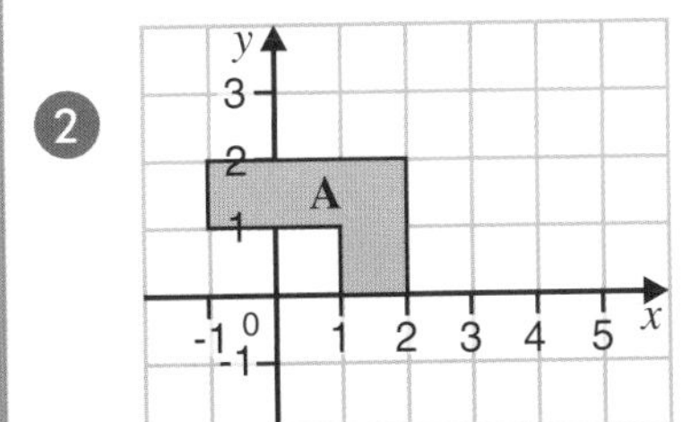

OCR

3 A is mapped onto B by a single transformation.
Describe the transformation.

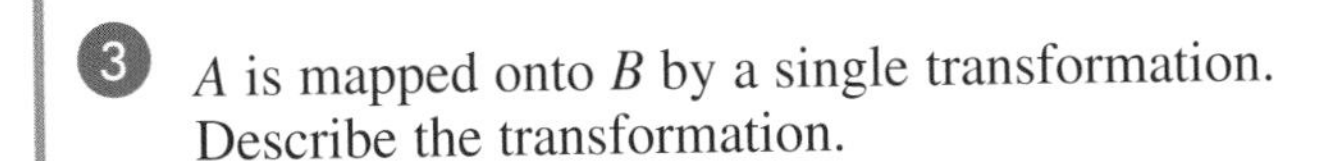

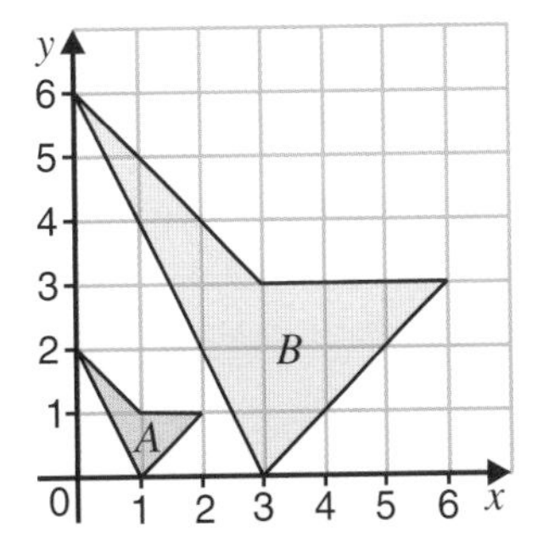

4
(a) P is mapped onto S by an enlargement.
What is the centre and scale factor of the enlargement?
(b) Copy shape P onto squared paper.
Draw an enlargement of shape P with scale factor 2, centre (3, 2).

5 Copy triangle T onto squared paper.
Enlarge triangle T by scale factor $\frac{1}{3}$ with centre the point $P(1, 4)$.

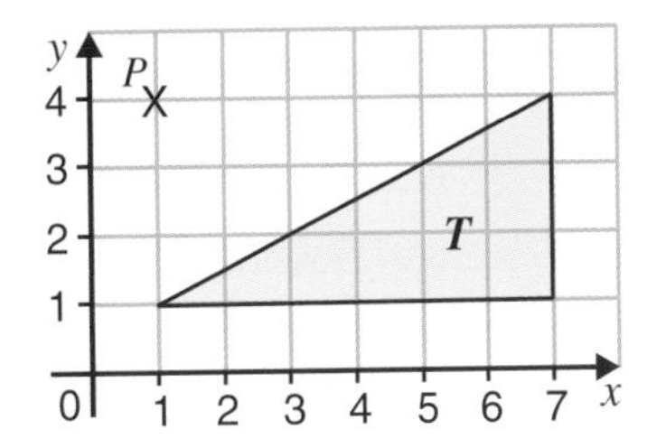

6 The diagram shows rectangles **A**, **B** and **C**.

A: 2 cm × 5 cm	**B**: 3 cm	**C**: 4 cm × 7 cm

(a) Explain why rectangles **A** and **C** are **not** similar.
(b) Rectangles **A** and **B** are similar. Work out the length of rectangle **B**.

7 ABC and PQR are similar triangles.
Find the length marked (a) x
(b) y

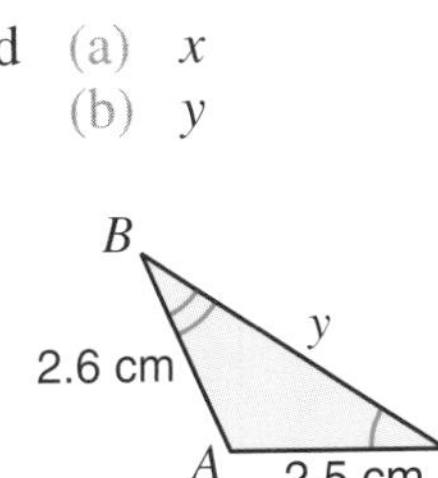

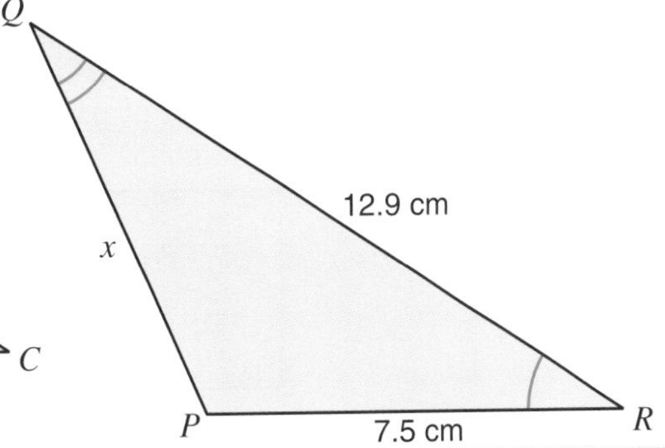

OCR

SECTION 32 Pythagoras' Theorem

What you need to know

- The longest side in a right-angled triangle is called the **hypotenuse**.
- The **Theorem of Pythagoras** states:
  "In any right-angled triangle the square on the hypotenuse is equal to the sum of the squares on the other two sides."
  $a^2 = b^2 + c^2$

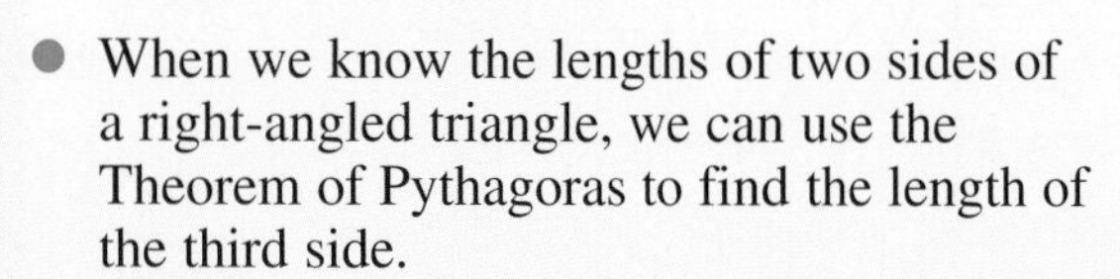

- When we know the lengths of two sides of a right-angled triangle, we can use the Theorem of Pythagoras to find the length of the third side.

$a^2 = b^2 + c^2$
Rearranging gives: $b^2 = a^2 - c^2$
$c^2 = a^2 - b^2$

Eg 1 Calculate the length of side a.

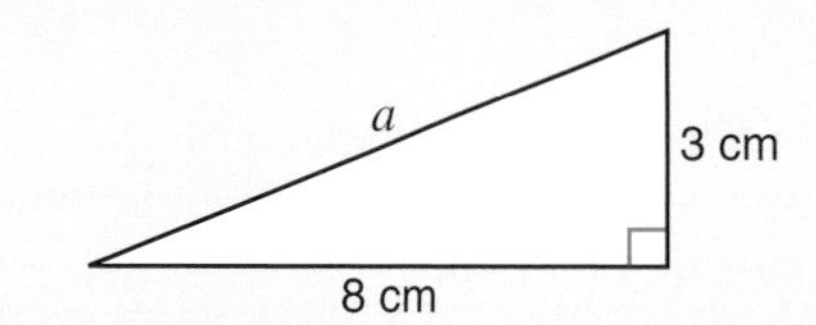

$a^2 = b^2 + c^2$
$a^2 = 8^2 + 3^2$
$a^2 = 64 + 9 = 73$
$a = \sqrt{73} = 8.544...$
$a = 8.5$ cm, correct to 1 d.p.

Eg 2 Calculate the length of side b.

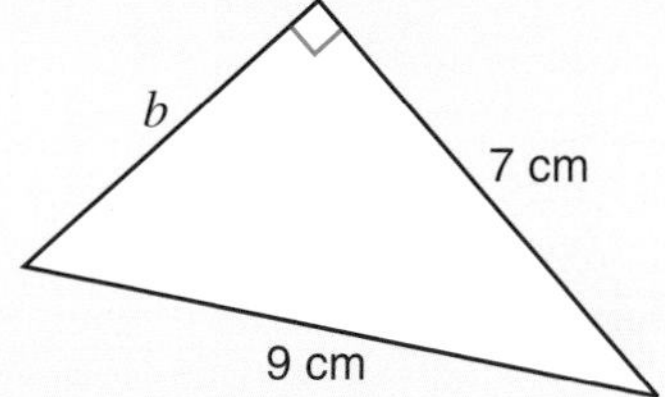

$b^2 = a^2 - c^2$
$b^2 = 9^2 - 7^2$
$b^2 = 81 - 49 = 32$
$b = \sqrt{32} = 5.656...$
$b = 5.7$ cm, correct to 1 d.p.

Exercise 32

Do not use a calculator for questions 1 and 2.

1. ABC is a right-angled triangle.
   $AB = 5$ cm and $AC = 12$ cm.
   Calculate the length of BC.

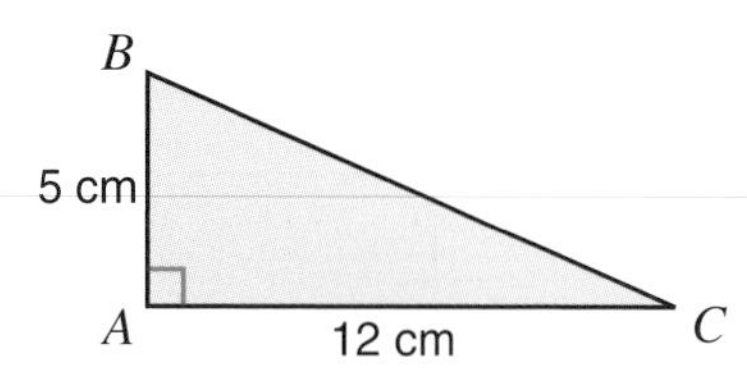

2. 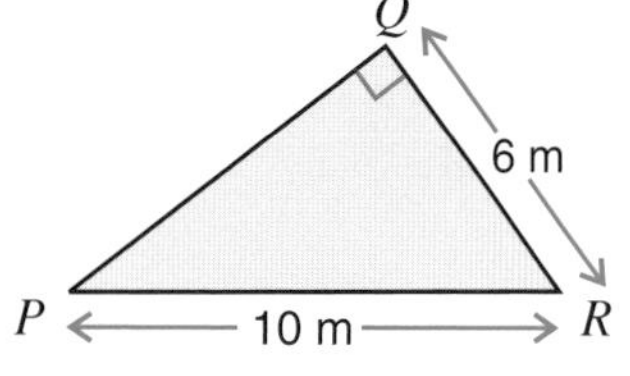

   The diagram shows the cross-section of the roof of a house.
   The width of the house, PR, is 10 m.
   $QR = 6$ m and angle $PQR = 90°$.
   Calculate the length of PQ.

3. The diagram shows a rectangular sheet of paper.
   The paper is 20 cm wide and the diagonal, d, is 35 cm.

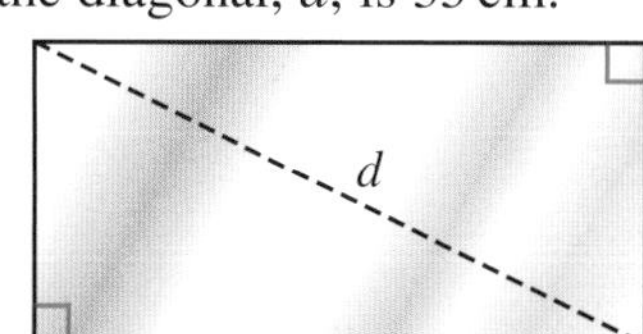

   Calculate the length of the sheet of paper.

4 The diagram shows the points $A(1, 2)$ and $B(5, 8)$.

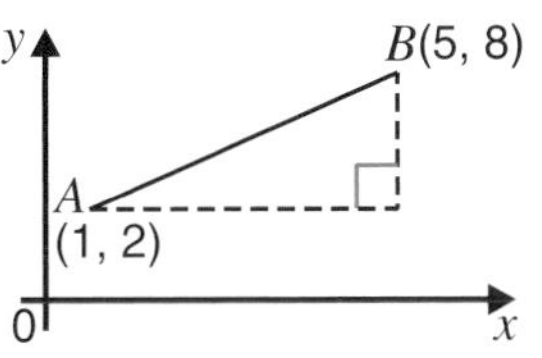

(a) Find the coordinates of the midpoint of AB.
(b) Calculate the length of the line AB. OCR

5 A football pitch measures 100 m by 72 m.
Mike walks along the edge of the pitch from A to B.
Alan walks diagonally across the pitch from A to B.

Calculate how much further Mike walks than Alan.

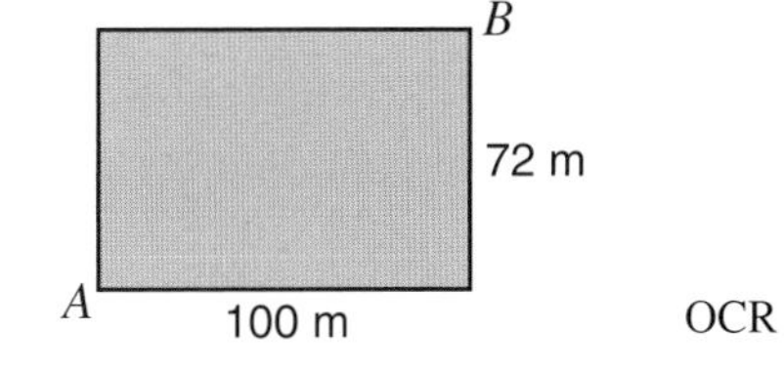

OCR

6 PQR is a right-angled triangle. $PQ = 5$ cm and $PR = 9$ cm.

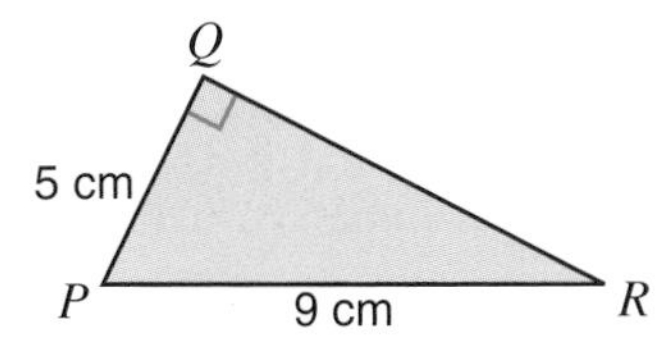

Calculate the length of QR and, hence, find the area of triangle PQR.

7 The diagram represents the frame of a slide in which ACD is horizontal and BC is vertical.

$AC = 5.1$ m, $BC = 2.3$ m and $BD = 3.8$ m.

Calculate the length of AB.

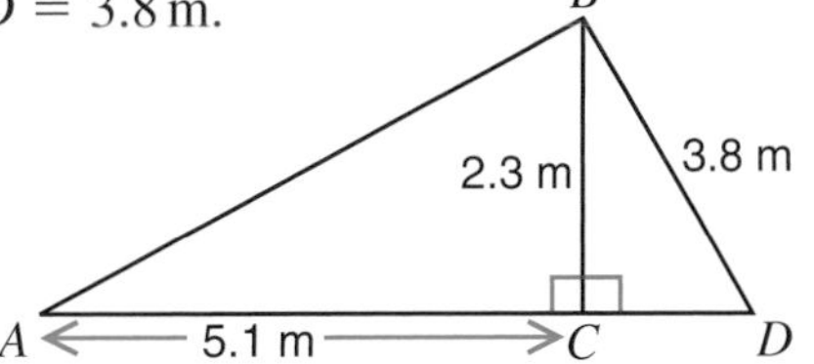

OCR

8

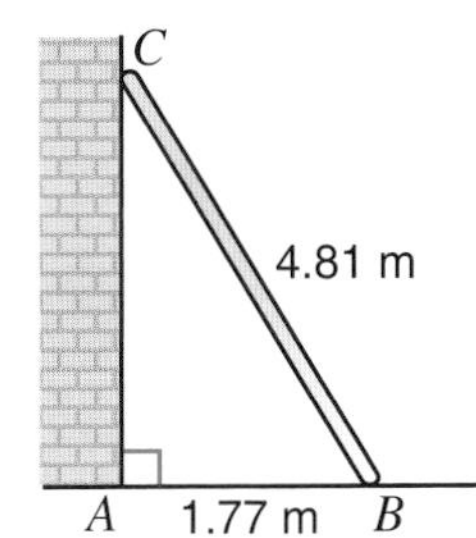

A ladder, BC, leans against a vertical wall.
The bottom of the ladder rests on horizontal ground.

$BC = 4.81$ m and $AB = 1.77$ m.

Calculate the distance AC.
Give your answer to a sensible degree of accuracy. OCR

9 The square, $ABCD$, has an area of 50 cm^2.

Calculate the length of the diagonal, AC.

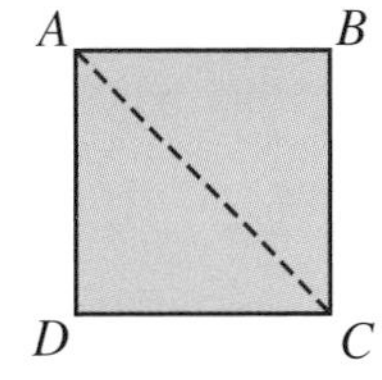

10 $ABCD$ is a rectangle. $AD = 5$ cm, $DC = 9$ cm and $EC = 6$ cm.

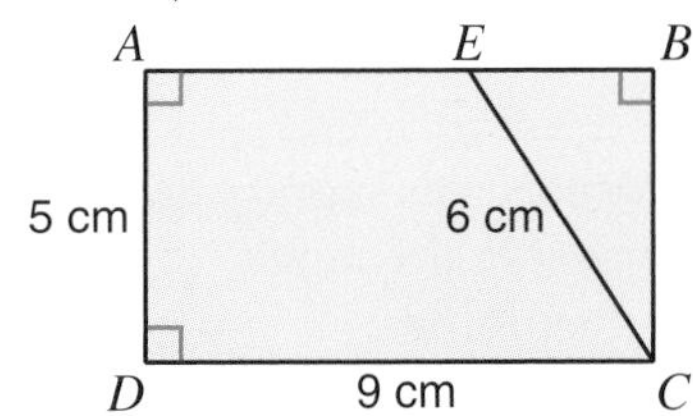

Calculate the length of AE, correct to one decimal place.

SECTION 33

Understanding and Using Measures

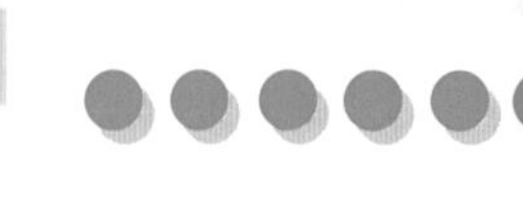

What you need to know

- The common units – both **metric** and **imperial** – used to measure **length**, **mass** and **capacity**.
- How to estimate measurements using sensible units and a suitable degree of accuracy.
- How to convert from one unit to another. This includes knowing the connection between one metric unit and another and the approximate equivalents between metric and imperial units.

Metric Units

Length
1 kilometre (km) = 1000 metres (m)
1 m = 100 centimetres (cm)
1 cm = 10 millimetres (mm)

Mass
1 tonne (t) = 1000 kilograms (kg)
1 kg = 1000 grams (g)

Capacity and volume
1 litre = 1000 millilitres (ml)
$1 \text{ cm}^3 = 1 \text{ ml}$

Imperial Units

Length
1 foot = 12 inches
1 yard = 3 feet

Mass
1 pound = 16 ounces
14 pounds = 1 stone

Capacity and volume
1 gallon = 8 pints

Conversions

Length
5 miles is about 8 km
1 inch is about 2.5 cm
1 foot is about 30 cm

Mass
1 kg is about 2.2 pounds

Capacity and volume
1 litre is about 1.75 pints
1 gallon is about 4.5 litres

- How to change between units of area. For example $1 \text{ m}^2 = 10\,000 \text{ cm}^2$.
- How to change between units of volume. For example $1 \text{ m}^3 = 1\,000\,000 \text{ cm}^3$.
- You should be able to solve problems involving different units.

Eg 1 A tank holds 6 gallons of water.
How many litres is this? $6 \times 4.5 = 27$ litres

Eg 2 A cuboid measures 1.5 m by 90 cm by 80 cm.
Calculate the volume of the cuboid, in m^3. $1.5 \times 0.9 \times 0.8 = 1.08 \text{ m}^3$

- Be able to read scales accurately.

Eg 3 Part of a scale is shown.
It measures weight in grams.
What weight is shown by the arrow?

The arrow shows 27 grams.

- Be able to recognise limitations on the accuracy of measurements.
A measurement given to the nearest whole unit may be inaccurate by one half of a unit in either direction.

Eg 4 A road is 400 m long, to the nearest 10 m.
Between what lengths is the actual length of the road?
Actual length = 400 m ± 5 m $\quad 395 \text{ m} \leq \text{actual length} < 405 \text{ m}$

- By analysing the **dimensions** of a formula it is possible to decide whether a given formula represents a **length** (dimension 1), an **area** (dimension 2) or a **volume** (dimension 3).

Eg 5 p, q, r and s represent lengths.
By using dimensions, decide whether the expression $pq + qr + rs$ could represent a perimeter, an area or a volume.
Writing $pq + qr + rs$ using dimensions:

$$L \times L + L \times L + L \times L = L^2 + L^2 + L^2 = 3L^2$$

So, $pq + qr + rs$ has dimension 2 and could represent an area.

Exercise 33

Do not use a calculator for questions 1 to 6.

1 What value is shown by the pointer on each of these diagrams?

(a)

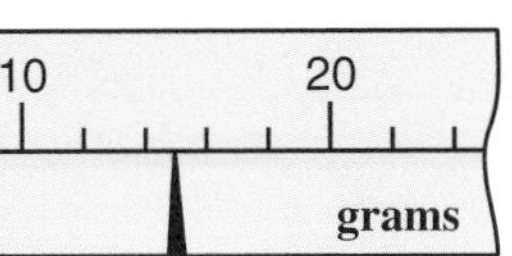

(b)

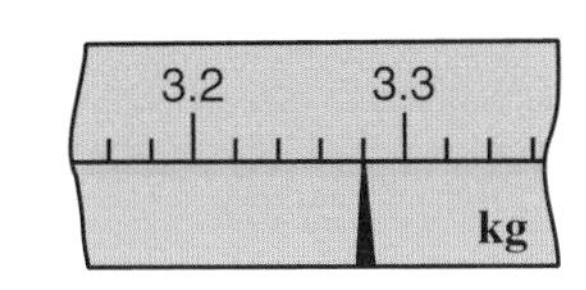

(c)

2 What metric unit is used to measure each of the following?
(a) the area of a carpet,
(b) the weight of an egg,
(c) the distance travelled on a train journey,
(d) the amount of petrol in the fuel tank of a car. OCR

3 Put these four lengths in order, shortest first.

2 mm **2 m** **20 cm** **20 mm** OCR

4 A glass contains 250 ml of milk.
What fraction of a litre is this?

5 (a)

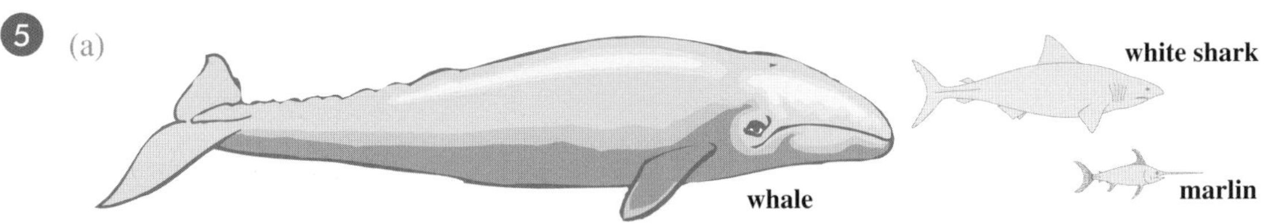

The length of the white shark is 9 m.
(i) Estimate the length of the whale.
(ii) Estimate the length of the marlin.

(b) In an angling contest the weight of the largest fish caught was 1.8 kg.
Write 1.8 kg in grams. OCR

6 The largest newspaper sold had a page size 1.4 m by 1 m.
The area of each page was 1.4 square metres
Work out this area in square centimetres. OCR

7 (a) A rectangular doormat measures 150 cm by 120 cm.
Calculate the area of the doormat in square metres.
(b) Change $0.2\,\text{m}^3$ to cm^3. OCR

8 Last year Felicity drove 2760 miles on business.
Her car does 38 miles per gallon. Petrol costs 89 pence per litre.
She is given a car allowance of 25 pence per kilometre.
How much of her car allowance is left after paying for her petrol?
Give your answer to the nearest £.

9 The length of Andy's pencil is 170 mm, correct to the nearest 10 mm.
What is the minimum length of the pencil?

10 A bag of carrots weighs 2.5 kg, correct to the nearest 100 g.
What is the minimum weight of the bag of carrots?

11 Pete cycles at 15 miles per hour.
What is his speed in kilometres per hour?

12 In these expressions, a, b and c represent lengths.

$$\pi(a+b) \qquad a^2 + ab + abc \qquad \frac{\pi a^2}{4} + \frac{\pi ac}{2} \qquad \pi a^2(b+c)$$

Which one of these expressions could represent an area? Show how you decide. OCR

Shape, Space and Measures Non-calculator Paper

Do not use a calculator for this exercise.

(a) Measure the line AB.

A ———————————————— B

Draw a copy of the line AB.

(b) Draw a circle with centre B and radius 3 cm.
(c) Draw a line from A, making an angle of 57° with AB.
(d) C is the midpoint of AB. Mark C on your diagram.
(e) Draw a line through C which is perpendicular to the line AB.
(f) Draw a line which is parallel to the line AB.

OCR

2 In which compass direction is
(a) the Bank from the Supermarket,
(b) the Supermarket from the Church?

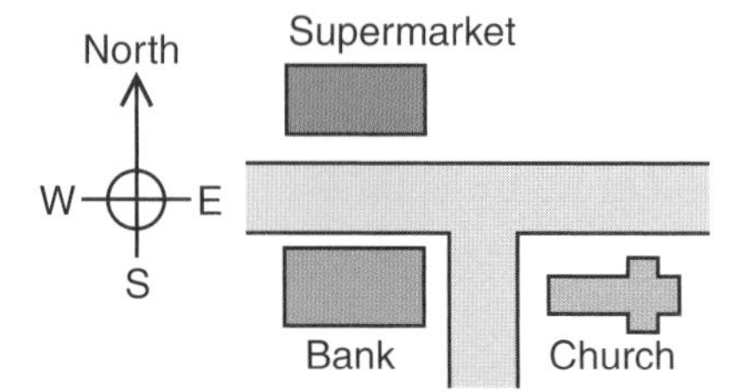

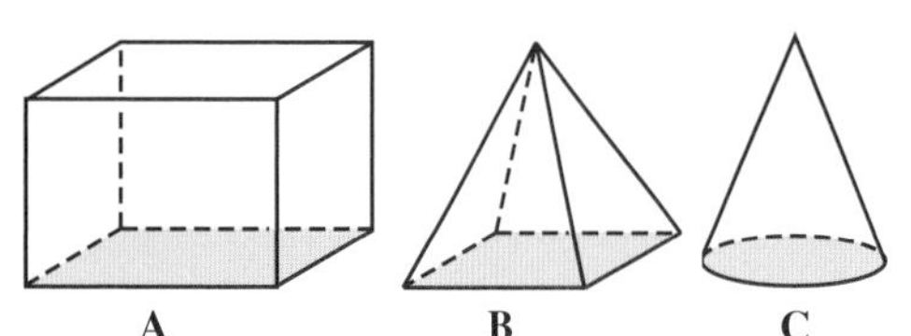

The diagram shows some 3-dimensional shapes.
(a) How many edges has shape **A**?
(b) How many faces has shape **B**?
(c) What is the mathematical name for shape **C**?

4 The shape has been drawn on 1 cm squared paper.
What is the area of the shape?

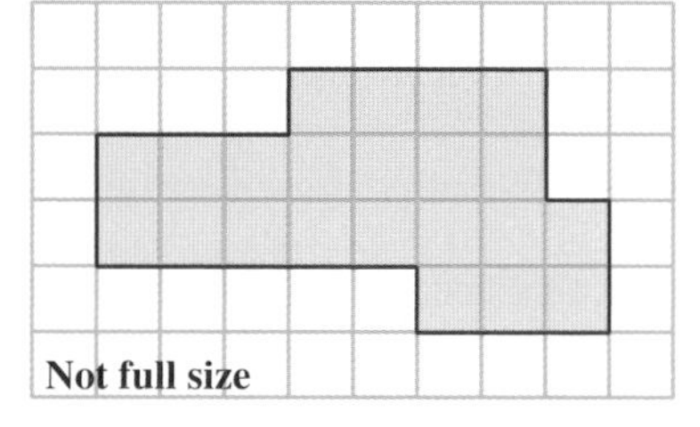

5

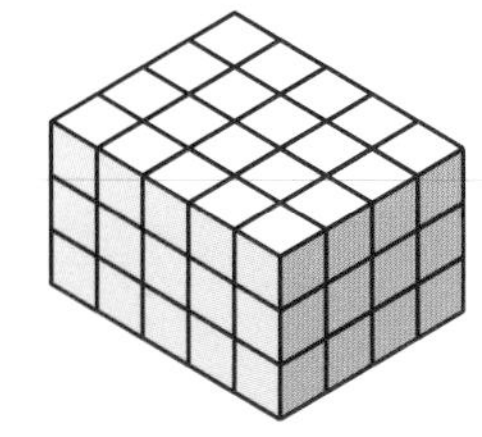

Stephen works in a supermarket.
He is stacking boxes.
(a) How many boxes are stacked here?

Each box is a cube of side 40 cm.
(b) What are the dimensions of his stack?
Give your answers in metres.

OCR

6 (a) What is the scale factor of the enlargement from **A** to **B**?
(b) When two quadrilaterals are enlargements of each other, like **A** and **B**, they are **similar**.
What mathematical word describes quadrilaterals which are identical in size and shape, like **A** and **C**?
(c) What is the area of (i) shape **A**, (ii) shape **B**?
(d) Describe the translation that maps **A** onto **C**.

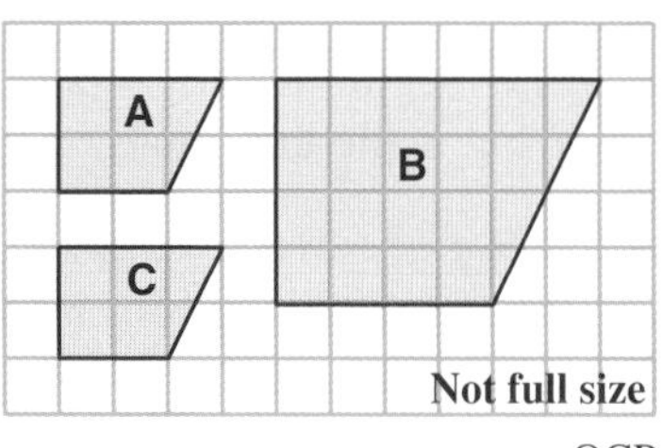

OCR

7 (a) Use your protractor to measure the size of angles x and y.
(b) Which of these angles is an obtuse angle?

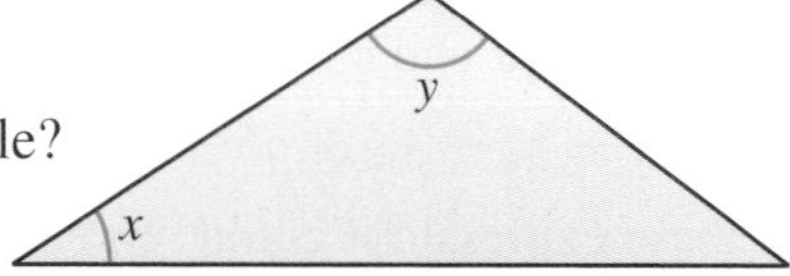

8 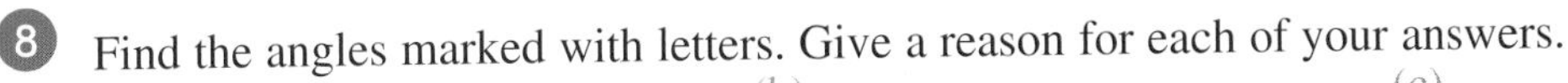
Find the angles marked with letters. Give a reason for each of your answers.

(a)

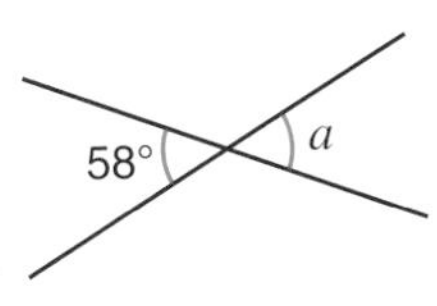

(b)

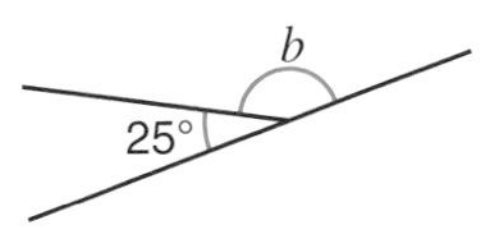

(c)

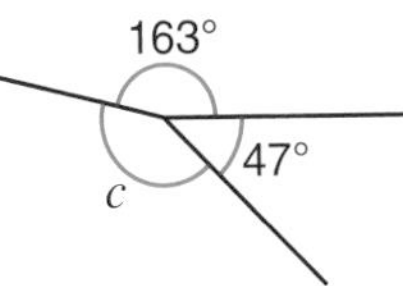

9 (a) Write the order of rotational symmetry for each shape.

(i)

(ii)

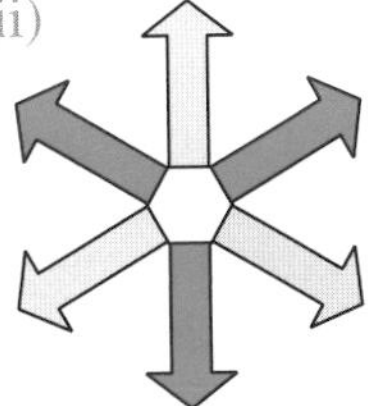

(iii)

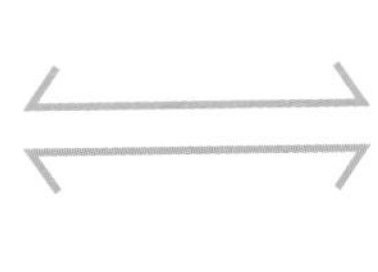

(b) On squared paper, draw a simple shape which has rotational symmetry of order 2, but has no lines of symmetry.

(c) Reflect this triangle in the line $x = 2$.

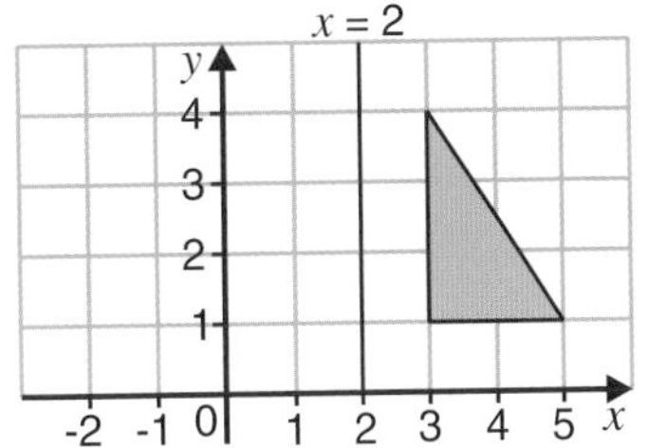

OCR

10

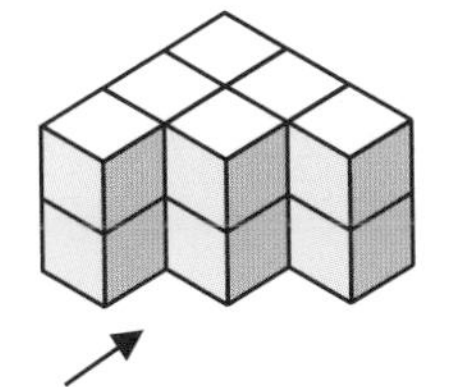

This solid has been made using 1 cm cubes.

(a) Draw the plan of the solid.

(b) Draw the elevation of the solid from the direction shown by the arrow.

11 Find the size of the angles a, b and c. Give a reason for each of your answers.

(a)

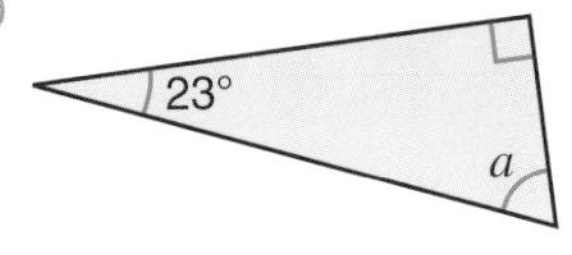

(b)

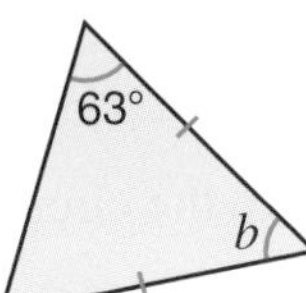

(c)

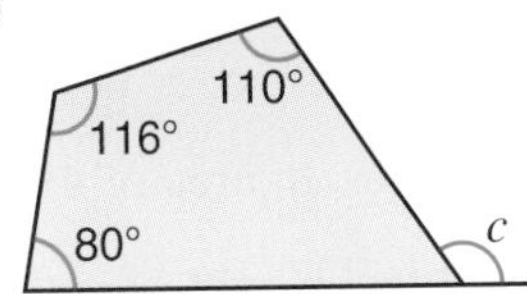

12 A charity asked people to donate 2p coins.
The 2p coins were laid out touching each other in a straight line.
The distance across each coin was 2.5 cm.
How long was a line of 1000 of the 2p coins?
Give your answer (a) in centimetres, (b) in metres.

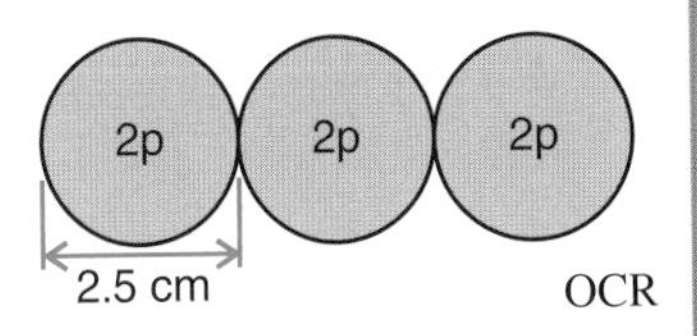

OCR

13

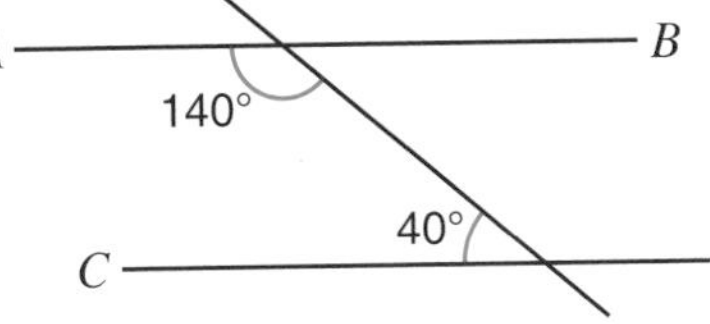

Give a reason why AB is parallel to CD.

OCR

14 The diagram shows the positions of three towns, A, B and C.

B is 16 km East of A. $AC = 10$ km and $BC = 13$ km.

(a) Make an accurate scale drawing of triangle ABC.
Use a scale of **1 cm to 2 km**.

(b) Use your drawing to find the bearing of C from B.

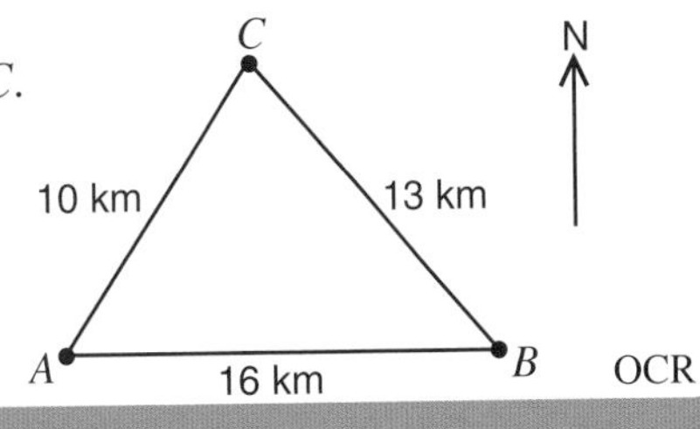

OCR

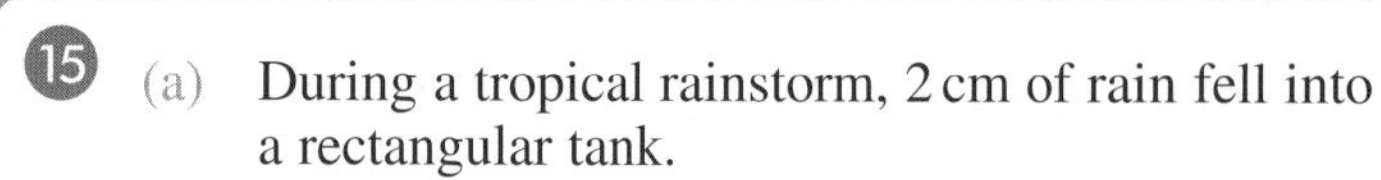

15 (a) During a tropical rainstorm, 2 cm of rain fell into a rectangular tank.
The base of the tank measures 40 cm by 30 cm.
(i) Calculate the volume of water in the tank.
(ii) Change your answer to part (a)(i) into litres.

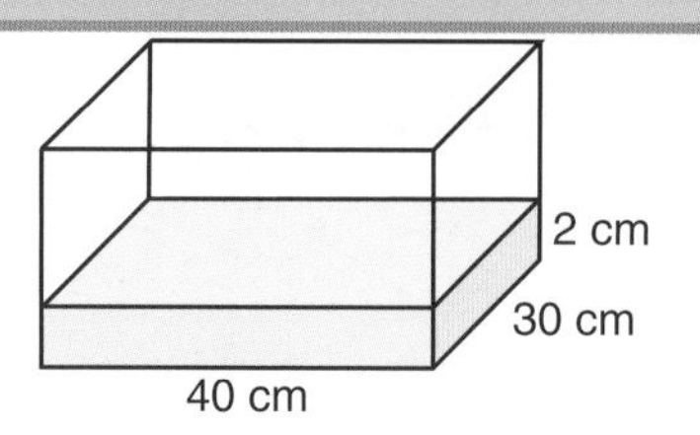

(b) On another day, 2000 cm^3 of water was collected in a different tank.
The base of this tank is a square of side 20 cm. Calculate the depth of the water in the tank.

OCR

16

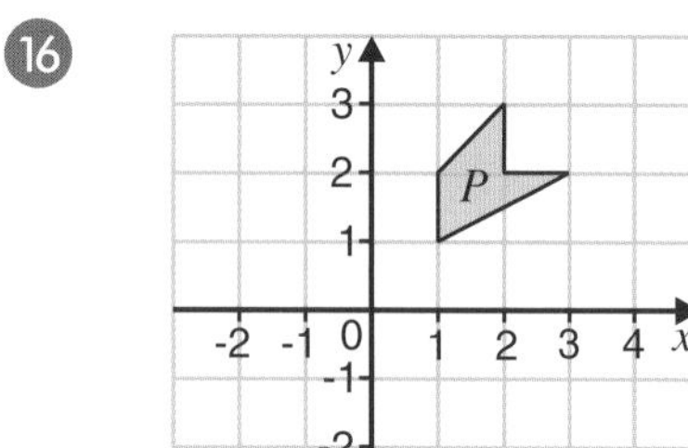

Copy the diagram onto squared paper.
(a) P is mapped onto Q by an enlargement, scale factor 2, centre $(-1, 3)$. Draw and label Q.
(b) P is mapped onto R by a translation with vector $\begin{pmatrix}-3\\2\end{pmatrix}$. Draw and label R.
(c) P is mapped onto S by a rotation through 90° clockwise, about $(1, 0)$. Draw and label S.

17 In the diagram, ACF is a straight line.
AD is parallel to CE.
$AC = BC$.
Calculate (a) x, (b) y, (c) z.

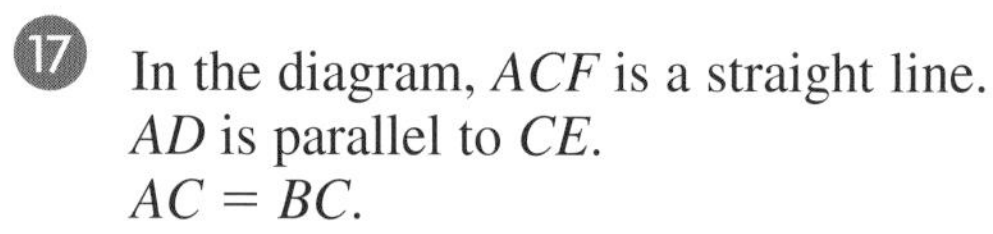

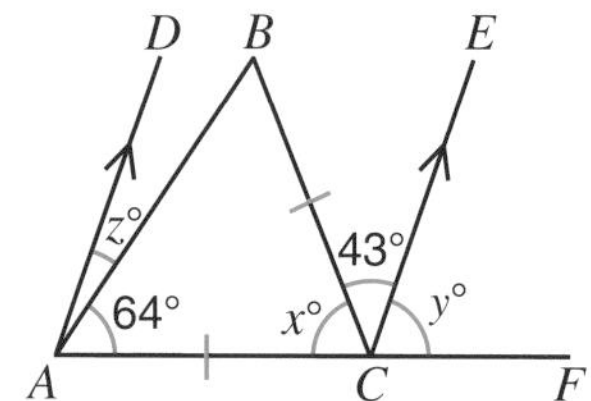

OCR

18 (a) Construct triangle ABC,
in which $AB = 9.5$ cm, $BC = 8$ cm and $CA = 6$ cm.
(b) Using ruler and compasses only, bisect angle BAC.
(c) Shade the region inside the triangle where all the points are less than 7.5 cm from B, and nearer to AC than to AB.

19 These two cars are similar.
Calculate h, the height, of the smaller car.

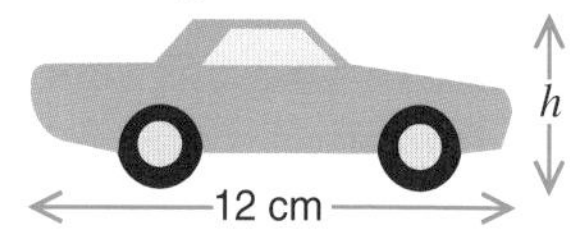

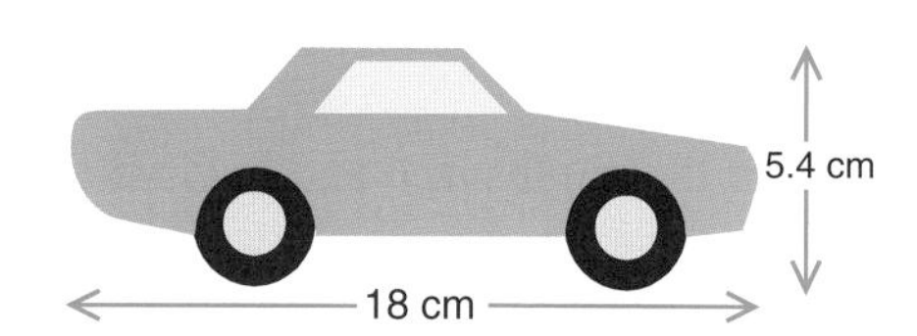

20 A circle has an area of $49\pi\text{ cm}^2$. Calculate the circumference of the circle in terms of π.

21 The vertical height of a slide is 2.3 m, correct to the nearest 0.1 m.
Write down the minimum possible value of the vertical height.

OCR

22 The following formulae represent certain quantities connected with containers, where a, b and c are dimensions.

πa $\quad abc$ $\quad \sqrt{a^2 - c^2}$ $\quad \pi a^2 b$ $\quad 2(a + b + c)$

(a) Explain why abc represents a volume.
(b) Which of these formulae represent lengths?

23 (a) The roof of a barn is a prism.
The cross-section is an isosceles triangle.
The dimensions, in metres, are given on the diagram.
(i) Show that $x = 5$.
(ii) Work out the total surface area of the four sections of the barn roof.

(b) The roof of another barn is a triangular prism, as shown in the diagram.
The dimensions, in metres, are given on the diagram.
The volume of the roof space is 48 m^3.
Work out the length of the roof, y metres.

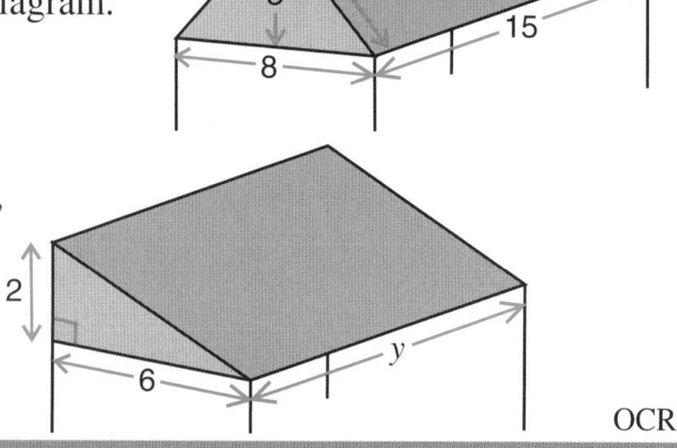

OCR

Shape, Space and Measures Calculator Paper

You may use a calculator for this exercise.

1 (a) Which of these weights are the same?
8000 g   80 kg   800 g   8 kg   0.08 kg

(b) Which of these lengths is the longest?
0.2 km   20 m   2000 mm   200 cm

(c) The scales show weights in kilograms.
Write down the weight of the pears.

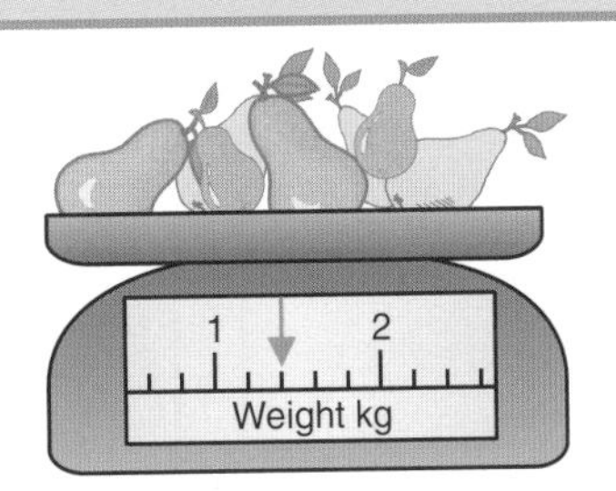

2 Look at these shapes.

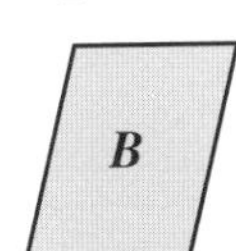

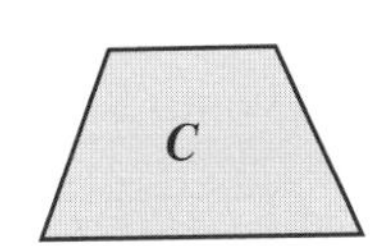

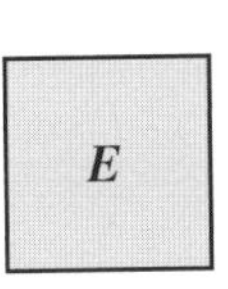

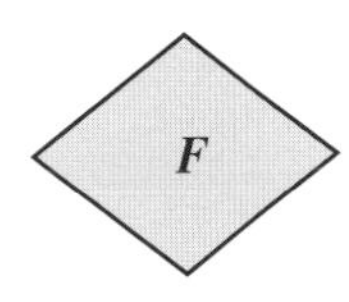

Write the letter of the correct shape for each of these names.

Square   **Trapezium**   **Parallelogram**   **Rhombus**

OCR

3 This shape is rotated through 180° about centre O.
Copy the diagram and draw the rotation.

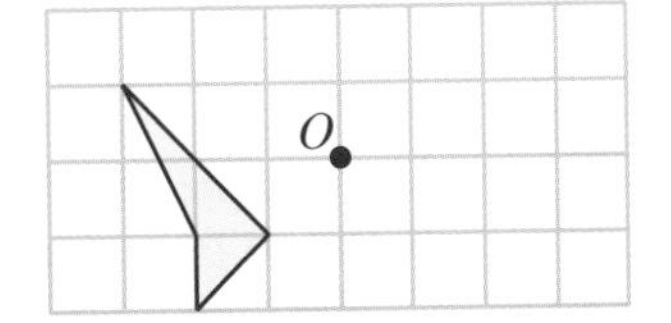

4

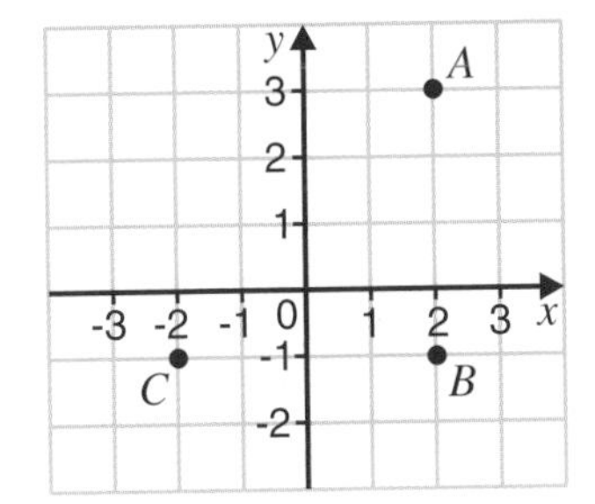

The diagram shows points A, B and C.

(a) What are the coordinates of A?

(b) What are the coordinates of C?

(c) $ABCD$ is a square.
What are the coordinates of D?

5 Here is a circle.

(a) Measure the diameter of the circle.

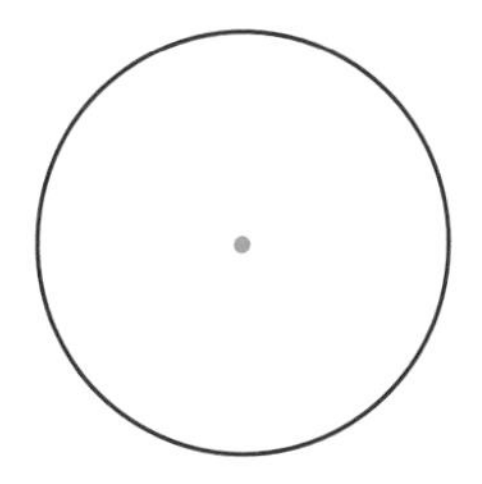

Draw a circle of your own.

(b) Mark a point on the circumference of your circle. Label the point C.

(c) On your diagram, draw a radius of the circle. Label it R.

(d) The circumference of a circle is given by this formula.

circumference = 3.14 × diameter

What is the circumference of a circle of diameter 100 cm?

OCR

6 Colin is 5 feet 10 inches tall and weighs 11 stones.
On a medical form he is asked to give his height in centimetres and his weight in kilograms.
What values should he give?

7 (a) What is the volume of the cuboid?

(b) What is the area of the shaded top of the cuboid?

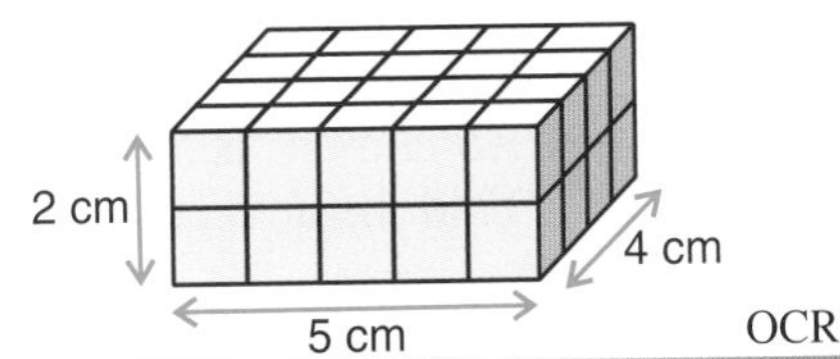

OCR

8 This diagram is wrong.
Explain why.

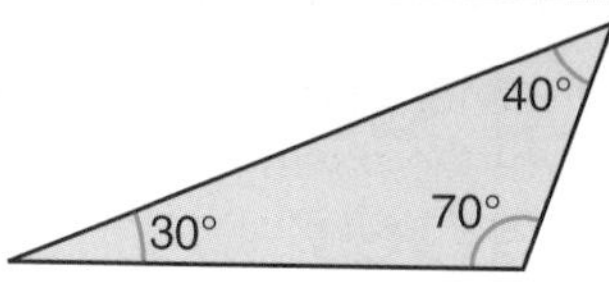

9 Write these volumes in order, smallest first. $\frac{1}{2}$ **litre** **2 litres** **200 ml** **0.7 litre** OCR

10 The diagram shows the positions of shapes P, Q, R and S.

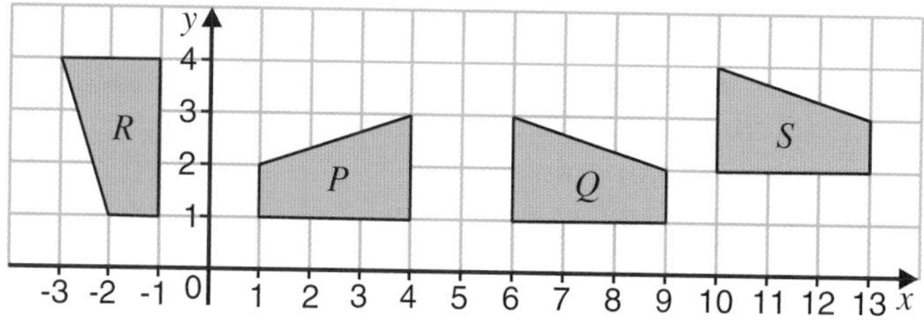

Describe the single transformation which takes:
(a) P onto Q, (b) P onto R, (c) Q onto S.

11 Work out the area of the shaded rectangle.

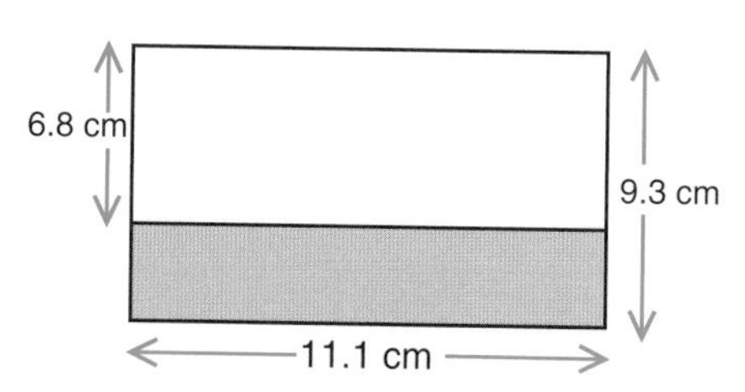

OCR

12

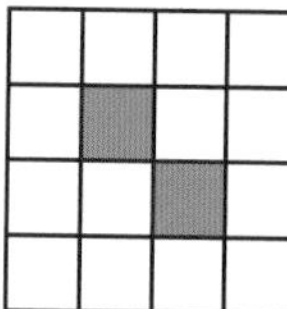

(a) Copy the diagram.
Shade two more squares so that the final diagram has line symmetry only.
(b) Make another copy of the diagram.
Shade two more squares so that the final diagram has rotational symmetry only.

13 (a) Part of a tessellation of triangles is shown.
Copy the diagram.
Continue the tessellation by drawing four more triangles.
(b) Do all regular polygons tessellate?
Give a reason for your answer.

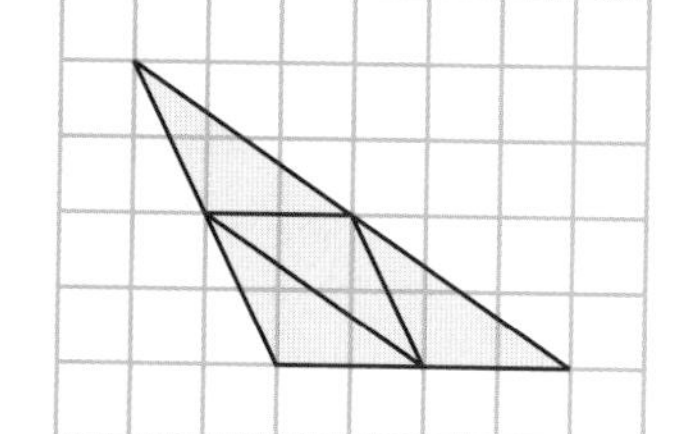

14 (a) A cuboid measures 2 cm by 2.5 cm by 4 cm.
(i) Draw an accurate net of the cuboid.
(ii) Calculate the total surface area of the cuboid.
(b) Another cuboid has a volume of 50 cm³. The base of the cuboid measures 4 cm by 5 cm.
Calculate the height of the cuboid.

15 (a) In the diagram, $AB = AC$, angle $A = 98°$
and BCD is a straight line.
(i) Find the value of x.
(ii) Find the value of y.
Give a reason for your answer.

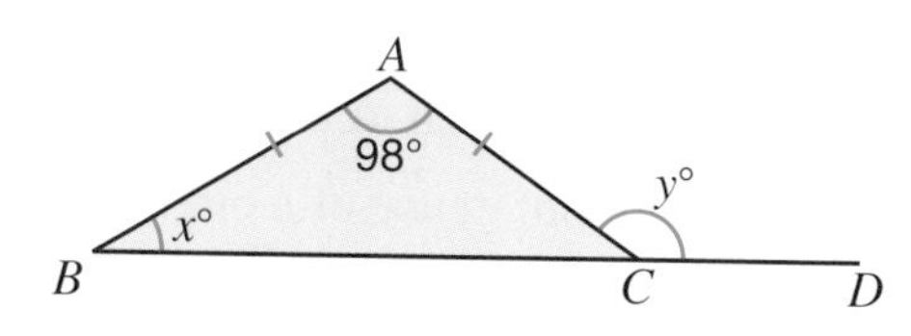

(b) $ABCD$ is a trapezium.
(i) Find the value of z.
Give a reason for your answer.
(ii) $AB = 24$ cm. $DC = 35$ cm.
The perpendicular distance between AB and DC is 5 cm.
Find the area of the trapezium $ABCD$.

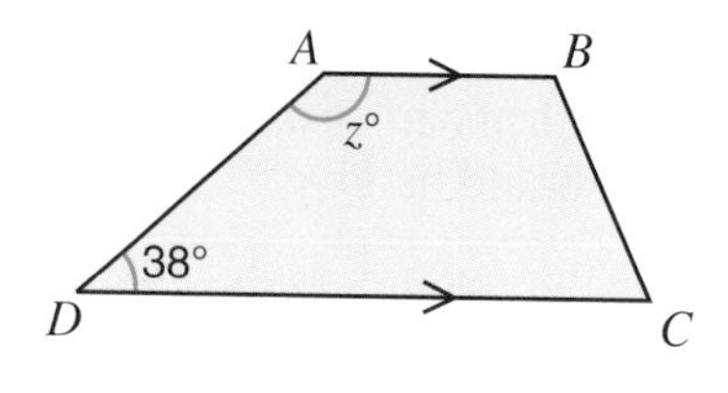

OCR

16 A circular plate has a diameter of 8 cm. Calculate the area of the plate.

17

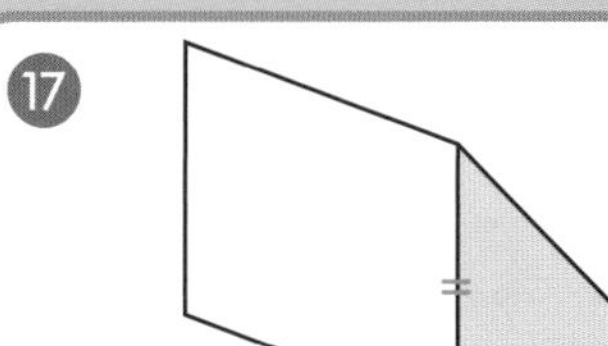

The diagram represents a prism.
The cross-section of the prism is an isosceles triangle.
Copy the diagram and draw one plane of symmetry of the prism.

18 The diagram shows the angle formed when three regular polygons are placed together, as shown.

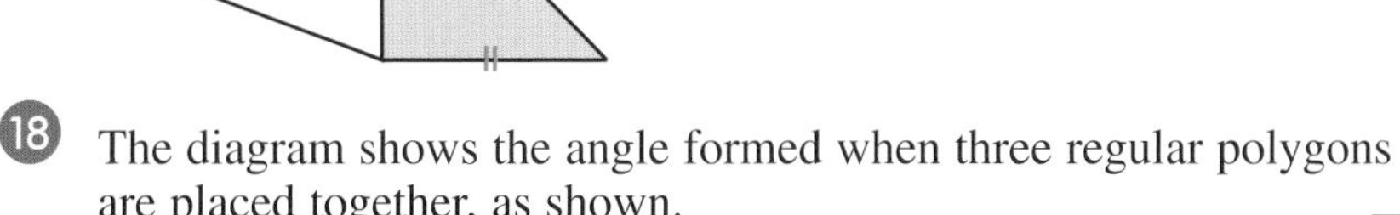

(a) Explain why angle a is 120°.
(b) Work out the size of the angle marked b.

19 On a map the distance between two hospitals is 14.5 cm.
The map has been drawn to a scale of 1 to 250 000.
Calculate the actual distance between the hospitals in kilometres.

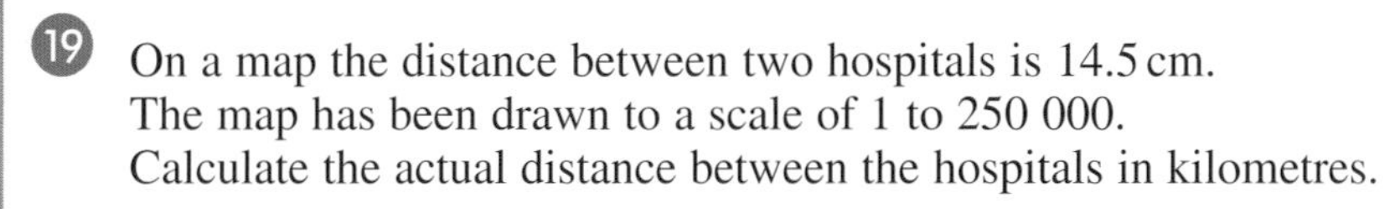

20

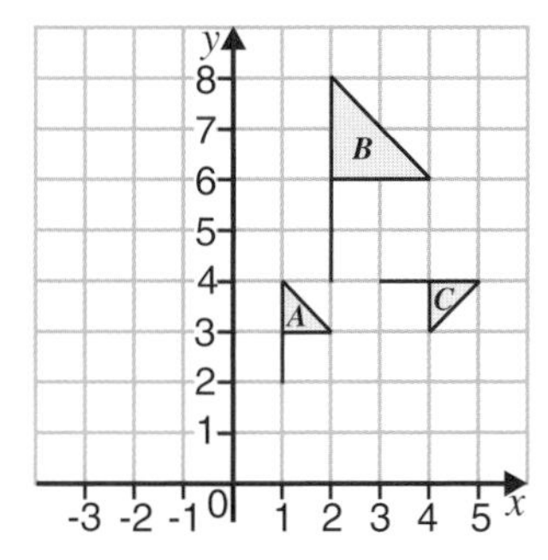

(a) Copy flag A onto squared paper.
Draw the image of the flag A after a translation $\begin{pmatrix} -4 \\ 3 \end{pmatrix}$.
Label it T.

(b) Describe fully the single transformation that will map
(i) A onto B, (ii) A onto C.

OCR

21 The diagram shows a plot of land.
The measurements are in metres.

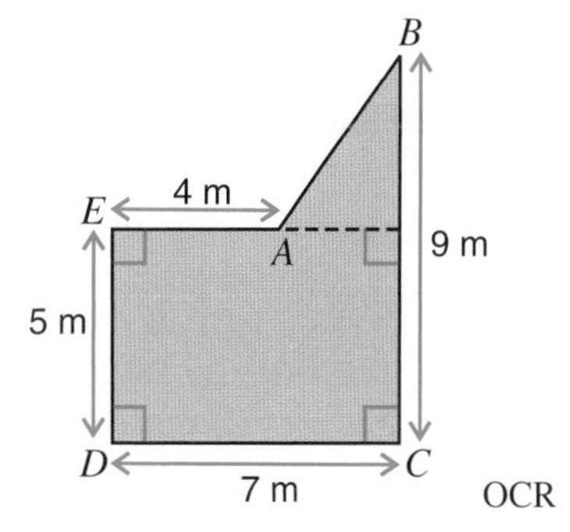

(a) The plot is to be turfed to make a lawn.
What will be the area of the lawn?

(b) (i) Find the length AB.
(ii) A fence is to be put around the perimeter of the plot.
What will be the length of the fence?

OCR

22 Three oil rigs, X, Y and Z, are supplied by boats from port P.

X is 15 km from P on a bearing of 050°. Y is 20 km from P on a bearing of 110°.
Z is equidistant from X and Y and 30 km from P.

(a) By using a scale of 1 cm to represent 5 km, draw an accurate diagram to show the positions of P, X, Y and Z.
(b) Use your diagram to find (i) the bearing of Y from Z,
(ii) the distance, in kilometres, of Y from Z.

23 The diagram shows Fay's house, H, and her school, S.
To get to school Fay has a choice of two routes.
She can either walk along Waverly Crescent
or along the footpaths HX and XS.
Waverly Crescent is a semi-circle with diameter 650 m.
The footpath HX is 250 m and meets the footpath XS at right-angles.
Which of these routes is shorter? By how much?

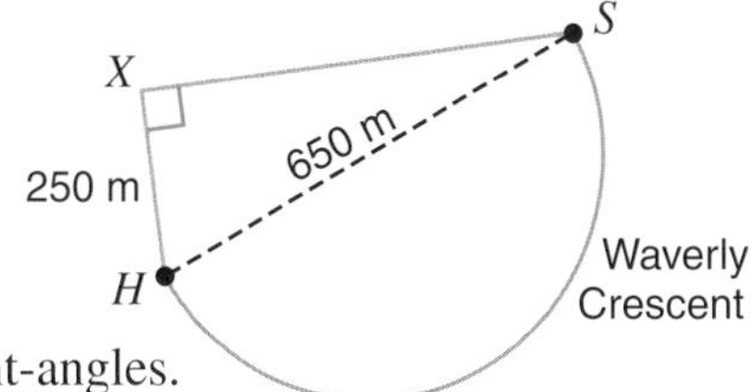

24

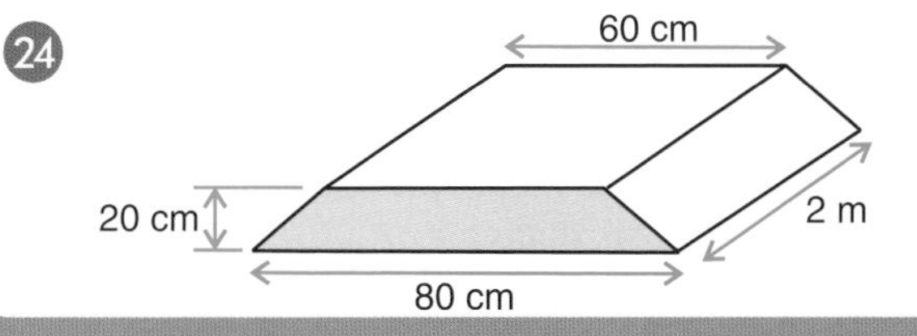

The diagram shows a block of wood which is 2 metres long.
The block is a prism with cross-section in the shape of a trapezium.
Find the volume of the block.

OCR

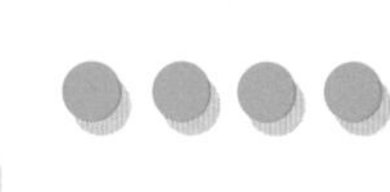

SECTION 34 Collection and Organisation of Data

What you need to know

- **Primary data** is data collected by an individual or organisation to use for a particular purpose.
  Primary data is obtained from experiments, investigations, surveys and by using questionnaires.
- **Secondary data** is data which is already available or has been collected by someone else for a different purpose.
  Sources of secondary data include the Annual Abstract of Statistics, Social Trends and the Internet.
- **Qualitative** data – Data which can only be described in words. E.g. Colour of cars.
- **Quantitative** data – Data that has a numerical value.
  Quantitative data is either **discrete** or **continuous**.
  Discrete data can only take certain values. E.g. Numbers of cars in car parks.
  Continuous data has no exact value and is measurable. E.g. Weights of cars.
- **Data Collection Sheets** – Used to record data during a survey.
- **Tally** – A way of recording each item of data on a data collection sheet.
  A group of five is recorded as ~~||||~~.
- **Frequency Table** – A way of collating the information recorded on a data collection sheet.
- **Grouped Frequency Table** – Used for continuous data or for discrete data when a lot of data has to be recorded.
- **Database** – A collection of data.
- **Class Interval** – The width of the groups used in a grouped frequency distribution.
- **Questionnaire** – A set of questions used to collect data for a survey.
  Questionnaires should:
  (1) use simple language,
  (2) ask short questions which can be answered precisely,
  (3) provide tick boxes,
  (4) avoid open-ended questions,
  (5) avoid leading questions,
  (6) ask questions in a logical order.
- **Hypothesis** – A hypothesis is a statement which may or may not be true.
- When information is required about a large group of people it is not always possible to survey everyone and only a **sample** may be asked.
  The sample chosen should be large enough to make the results meaningful and representative of the whole group (population) or the results may be **biased**.
- **Two-way Tables** – A way of illustrating two features of a survey.

Exercise 34

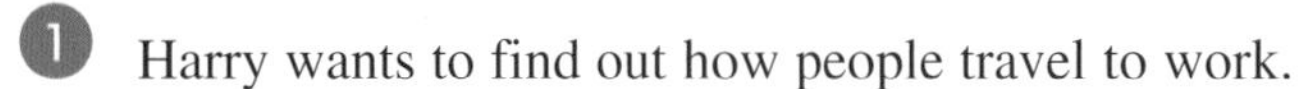

1 Harry wants to find out how people travel to work.
(a) (i) Design an observation sheet for Harry to record data.
(ii) Complete your observation sheet by inventing data for 20 people.
(b) Harry decides to stand outside the bus station to collect his data.
Give a reason why this is not a suitable place to carry out the survey.

2 The table shows information about pupils in the same class at a school.

Name	Gender	Month of birth	Day of birth
Corrin	F	June	Monday
Daniel	M	March	Thursday
Laila	F	May	Friday
Ria	F	March	Tuesday
Miles	M	April	Tuesday

(a) Who was born in May?
(b) Who was born on a Tuesday in March?
(c) Which of these pupils is most likely to be the youngest? Give a reason for your answer.

3 Tayfan is organising a skiing holiday to Italy for his friends.
They can go to Cervinia, Livigno or Tonale.
He asks each of his friends which resort they would like to go to and records the answers in his notebook.

Cervinia	Cervinia	Livigno	Tonale
Tonale	Tonale	Livigno	Cervinia
Livigno	Cervinia	Tonale	Tonale
Cervinia	Livigno	Tonale	Livigno
Tonale	Cervinia	Livigno	Tonale
Livigno	Tonale	Cervinia	

Show a better way of recording this information.

4 Dalbir conducted a survey on absence in his school.
He recorded the number of absences for each student for a term.
Here are the results for his form.

0	5	26	8	15	1	20	10
29	2	6	22	17	21	13	5
7	16	0	11	6	3	0	18
18	14	24	4	8	0	19	23

(a) Copy and complete this frequency table for the results.

Number of absences	Tally	Frequency
0 - 4		
5 - 9		
10 - 14		

(b) (i) How many students were there in his form?
(ii) How many of these students had fewer than 20 absences? OCR

5 The students in Aimee's class walk, cycle or catch the bus to school.
There are 30 students in her class and 13 of the students are boys.
8 of the boys cycle to school but none of the girls cycle.
9 of the girls walk to school. 12 students catch the bus to school.
(a) Copy and complete the table.

	Walk	Bus	Cycle	Totals
Boys				
Girls				
Totals				

(b) How many boys in Aimee's class walk to school?

6 Jamie is investigating the use made of his college library. Here is part of his questionnaire:

Library Questionnaire

1. How old are you? ....

(a) (i) Give a reason why this question is unsuitable.
(ii) Rewrite the question so that it could be included.

(b) Jamie asks the librarian to give the questionnaires to students when they borrow books.
(i) Give reasons why this sample may be biased.
(ii) Suggest a better way of giving out the questionnaires.

7 Julie is writing questions for a survey. Her hypothesis is that the main reason why people live in the Sharrow area is the cost of housing.
This is one of her questions.

Do you agree that the cost of housing is the main reason that people live in the Sharrow area, and if not, what do you think is the main reason?

State two things wrong with this question. OCR

8 This sample was used to investigate the claim:

"Women do more exercise than men."

	Age (years)			
	16 to 21	22 to 45	46 to 65	Over 65
Male	5	5	13	7
Female	25	35	0	0

Give three reasons why the sample is biased.

9 The table shows the results of a survey of 500 people.

	Can drive	Cannot drive
Men	180	20
Women	240	60

A newspaper headline states:

Survey shows that more women can drive than men.

Do the results of the survey support this headline?
Give a reason for your answer.

10 The two-way table shows the results of a survey of the number of cats and the number of dogs people have as pets.

Number of dogs \ Number of cats	0	1	2	3
0	21	9	3	0
1	5	4	2	0
2	2	1	0	1
3	1	1	0	0

(a) How many people have one dog **and** two cats as pets?

(b) A magazine article stated,

"Cats are more popular than dogs as pets."

Does this survey support that claim?
Give a reason for your answer.

(c) How many dogs did these people have altogether?

Pictograms and Bar Charts

- **Pictogram**. Symbols are used to represent information.
  Each symbol can represent one or more items of data.

Eg 1 A sports club has 45 members.
Last Saturday, 15 played football, 13 played hockey and 17 played rugby.
Draw a pictogram to show this information. Use [symbol] = 5 members.

Football	
Hockey	
Rugby	

Note: [symbol] represents 3 members
[symbol] represents 2 members

- **Bar chart**. Used for data which can be counted.
  Often used to compare quantities of data in a distribution.
  The length of each bar represents frequency.
  The longest bar represents the **mode**.
  The difference between the largest and smallest variable is called the **range**.

Bars can be drawn horizontally or vertically.
Bars are the same width and there are gaps between bars.

- **Bar-line graph**. Instead of drawing bars, horizontal or vertical lines are drawn to show frequency.

Eg 2 The graph shows the number of goals scored by a football team in 10 matches.

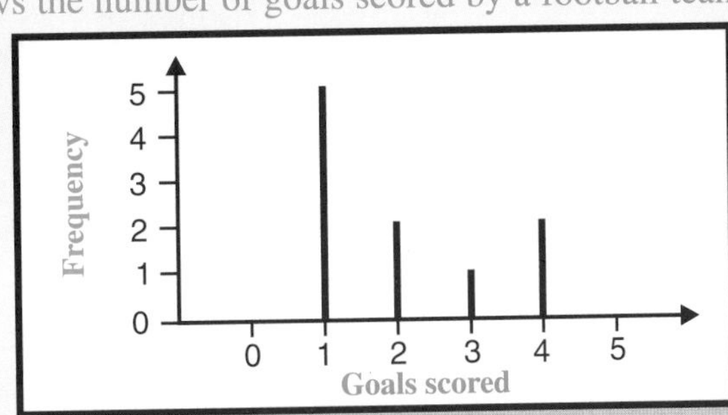

(a) Which number of goals scored is the mode?
(b) What is the range of the number of goals scored?

(a) The tallest bar represents the mode. The mode is 1 goal.
(b) The range is the difference between the largest and smallest number of goals scored.
The range $= 4 - 1 = 3$

Exercise 35

1 Caroline, Dolores, Fiona, Julie, Louise, Maureen, Natalie
Nicola, Pauline, Satinda

The vowels were taken from the ten girls' names above, as follows.

A, O, I, E, O, O, E, I, O, A, U, I, E, O, U, I, E, A
U, E, E, A, A, I, E, I, O, A, A, U, I, E, A, I, A

(a) Use a tally chart to record the number of times that each of the five vowels occurs.
(b) Show your results as a bar chart.
(c) What is the modal letter? OCR

35

2 The diagram shows the number of books borrowed from a school library last week.

Day	Symbols	Books
Monday		30
Tuesday		45
Wednesday		60
Thursday		

(a) How many books does ▢ represent?
(b) How many books were borrowed on Thursday?

50 books were borrowed on Friday.
(c) Draw the symbols to show the number of books borrowed on Friday. OCR

3 The bar-line graph shows the number of hours a plumber worked each day last week.

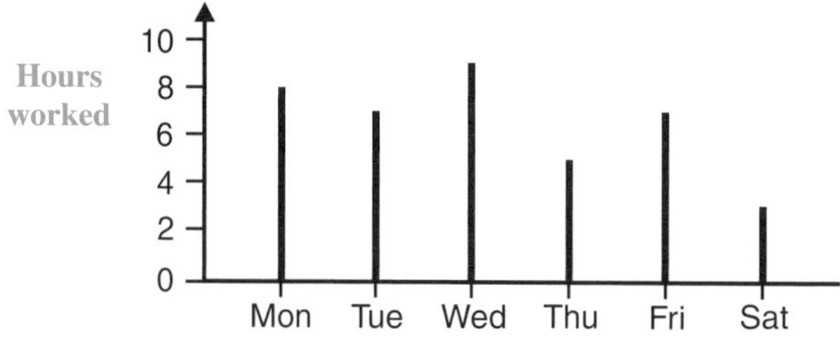

(a) On which day did he work the most hours?
(b) How many more hours did he work on Monday than on Thursday?
(c) How many hours did he work altogether last week?

4 Francis asks his friends to name their favourite flavour of yogurt.
The results are shown in the tally chart.

Flavour	Tally
Strawberry	𝍸 III
Vanilla	IIII
Other	𝍸 𝍸 III

(a) How many friends said strawberry?
(b) What percentage said vanilla?
(c) Draw a pictogram to show Francis's results.
Use the symbol 웃 to represent 5 friends.

5 Causeway Hockey Club have a hockey team for men and a hockey team for women.
The bar chart shows the number of goals scored in matches played by these teams last season.

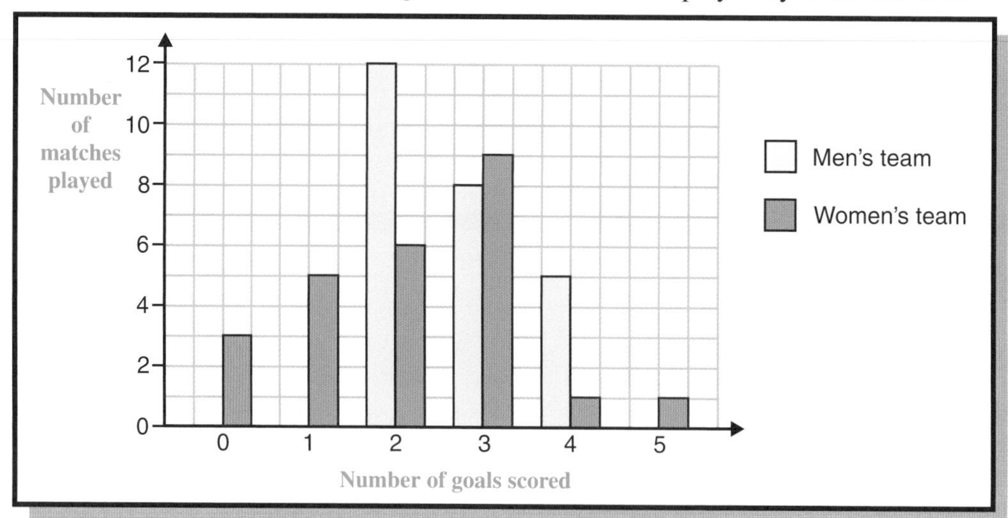

(a) For the men's team, find the range and mode in the number of goals scored.
(b) Compare and comment on the goals scored by these teams last season.

Averages and Range

What you need to know

- There are three types of **average**: the **mode**, the **median** and the **mean**.
  The **mode** is the most common value.
  The **median** is the middle value (or the mean of the two middle values) when the values are arranged in order of size.
  The **Mean** $= \dfrac{\text{Total of all values}}{\text{Number of values}}$
- The **range** is a measure of **spread**, and is the difference between the highest and lowest values.

Eg 1 The number of text messages received by 7 students on Saturday is shown.

2 4 3 4 4 3 2

Find (a) the mode, (b) the median, (c) the mean, (d) the range.

(a) The mode is 4.

(b) 2 2 3 (3) 4 4 4 The median is 3.

(c) The mean $= \dfrac{2+4+3+4+4+3+2}{7} = \dfrac{22}{7} = 3.14... = 3.1$, correct to 1 d.p.

(d) The range $= 4 - 2 = 2$

- To find the mean of a **frequency distribution** use:

$$\text{Mean} = \frac{\text{Total of all values}}{\text{Number of values}} = \frac{\Sigma fx}{\Sigma f}$$

Eg 2 The table shows the number of stamps on some parcels.

Number of stamps	1	2	3	4
Number of parcels	5	6	9	4

Find the mean number of stamps per parcel.

$$\text{Mean} = \frac{\text{Total number of stamps}}{\text{Number of parcels}} = \frac{1 \times 5 + 2 \times 6 + 3 \times 9 + 4 \times 4}{5 + 6 + 9 + 4} = \frac{60}{24} = 2.5$$

- To find the mean of a **grouped frequency distribution**, first find the value of the midpoint of each class.
  Then use:

$$\text{Estimated mean} = \frac{\text{Total of all values}}{\text{Number of values}} = \frac{\Sigma fx}{\Sigma f}$$

Eg 3 The table shows the weights of some parcels.

Weight (w grams)	Frequency
$100 \leq w < 200$	7
$200 \leq w < 300$	11
$300 \leq w < 400$	19
$400 \leq w < 500$	3

Calculate an estimate of the mean weight of these parcels.

$$\text{Mean} = \frac{\Sigma fx}{\Sigma f} = \frac{150 \times 7 + 250 \times 11 + 350 \times 19 + 450 \times 3}{7 + 11 + 19 + 3} = \frac{11\,800}{40} = 295 \text{ grams}$$

- You should be able to choose the best average to use in different situations:
  When the most **popular** value is wanted use the **mode**.
  When **half** of the values have to be above the average use the **median**.
  When a **typical** value is wanted use either the **mode** or the **median**.
  When all the **actual** values have to be taken into account use the **mean**.
  When the average should not be distorted by a few very small or very large values do **not** use the mean.

Exercise 36

Do not use a calculator for questions 1 to 3.

1 Nine students were asked to estimate the length of this line, correct to the nearest centimetre.

The estimates the students made are shown.

8 10 10 10 11 12 12 14 15

(a) What is the range in their estimates?
(b) Which estimate is the mode?
(c) Which estimate is the median?
(d) Work out the mean of their estimates.

2 Here are the number of eggs collected by a farmer over 10 days.

12, 12, 11, 13, 17, 19, 13, 12, 14, 17

(a) Find the median of these numbers.
(b) What is the range in the number of eggs collected?

3 The prices paid for eight different meals at a restaurant are:

£10 £9 £9.50 £12 £20 £11.50 £11 £9

(a) Which price is the mode?
(b) Find the median price.
(c) Calculate the mean price.
(d) Which of these averages best describes the average price paid for a meal? Give a reason for your answer.

4 A golfer played the eighteen holes on a course.
(a) His scores on the first nine holes are shown.

5 3 6 2 4 6 7 4 6

Find the range of these scores.
(b) His scores on the eighteen holes are shown in the table.

Score	2	3	4	5	6	7	8
Frequency	1	2	4	3	5	2	1

Write down the modal score.

OCR

5 (a) The number of hours of sunshine each day last week is shown.

Monday	Tuesday	Wednesday	Thursday	Friday	Saturday	Sunday
5.3	6.4	3.7	4.8	7.5	8.6	5.7

(i) What is the range in the number of hours of sunshine each day?
(ii) Work out the mean number of hours of sunshine each day.

(b) In the same week last year, the range in the number of hours of sunshine each day was 9 hours and the mean was 3.5 hours.
Compare the number of hours of sunshine each day in these two weeks.

6 The mean weight of the 8 Forwards in a rugby team is 108.25 kg.
The mean weight of the 7 Backs in the team is 89.5 kg.
Calculate the mean weight of the 15 members of the team.

OCR

7 Mansoor carried out a survey. He asked students how far they travel from home.
This table summarises his results.

	mean distance (km)	median distance (km)	range (km)
College	5.7	4.3	12
School	0.8	0.75	5.5

He makes these statements about his results. Complete his reasons.

(a) College students travel further than school students.
I know this because ....

(b) None of the school students travel more than 10 km, but at least one college student does.
I know this because .... OCR

8 Helen and Reg play ten-pin bowling. The graph shows their scores for the first 10 frames.

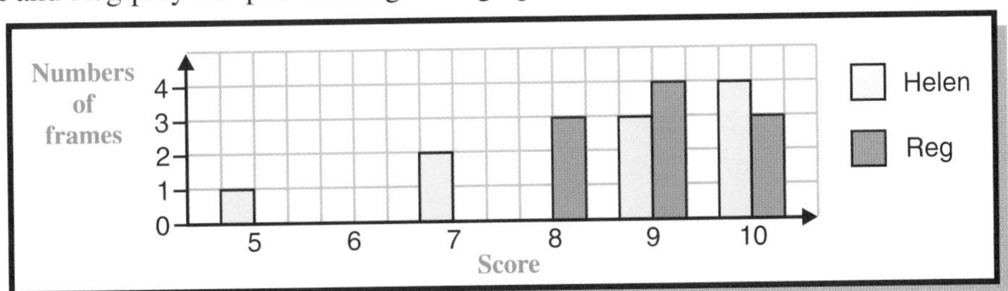

(a) What is the range in the scores for Helen?

(b) Find the mean of the scores for Reg.

(c) Reg says, "My average score is higher than Helen's."
Helen says, "My average score is higher than Reg's."
A friend says, "Your average scores are both the same."
Which average is being used by each person? Show your working.

9 Darren throws a dice 60 times. His results are shown.

Score	1	2	3	4	5	6
Frequency	12	10	9	11	10	8

(a) For these results, find
(i) the mode, (ii) the median, (iii) the mean.

(b) Darren throws the dice again and scores a 6.
Which of the averages he has found will not change?

10 Jenny asked 200 people how much they spent last year on magazines.
The results are in the table below.

Amount (£x)	$0 < x \leqslant 10$	$10 < x \leqslant 20$	$20 < x \leqslant 30$	$30 < x \leqslant 40$	$40 < x \leqslant 50$
Frequency	40	50	48	30	32

(a) Calculate an estimate of the mean amount spent on magazines.

(b) Explain briefly why this value of the mean is only an estimate. OCR

11 A manufacturer claims that the flavour of *Megagum* chewing gum lasts three times as long as any other gum. One hundred teenagers chew other brands of gum.
The mean time that the flavour lasts is calculated to be 2.08 hours.
The same 100 teenagers then chew *Megagum*.
The time that they each think the flavour lasts is shown in the grouped frequency table.

Time (t hours)	$5.8 < t \leqslant 5.9$	$5.9 < t \leqslant 6.0$	$6.0 < t \leqslant 6.1$	$6.1 < t \leqslant 6.2$	$6.2 < t \leqslant 6.3$
Frequency	2	15	47	34	2

(a) Calculate an estimate of the mean time that the flavour of *Megagum* lasts.

(b) Is the manufacturer's claim correct? Support your answer by calculation. OCR

SECTION 37 Pie Charts and Stem and Leaf Diagrams

What you need to know

- **Pie chart**. Used for data which can be counted.
  Often used to compare proportions of data, usually with the total.
  The whole circle represents all the data.
  The size of each sector represents the frequency of data in that sector.
  The largest sector represents the **mode**.

Eg 1 The pie chart shows the makes of 120 cars.
(a) Which make of car is the mode?
(b) How many of the cars are Ford?

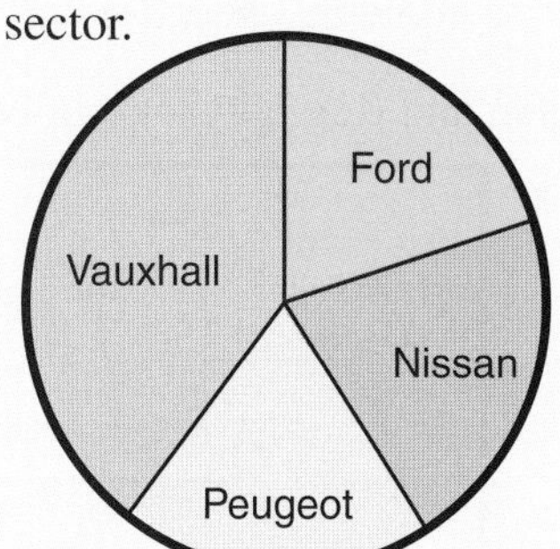

(a) The sector representing Vauxhall is the largest.
Therefore, Vauxhall is the mode.
(b) The angle of the sector representing Ford is 72°.
The number of Ford cars $= \frac{72}{360} \times 120 = 24$

- **Stem and leaf diagrams**. Used to represent data in its original form.
  Data is split into two parts.
  The part with the higher place value is the stem. E.g. 15 = stem 1, leaf 5.
  A key is given to show the value of the data. E.g. 3 | 4 means 3.4, etc.
  The data is shown in numerical order on the diagram. E.g. 2 | 3 5 9 represents 23, 25, 29.
  Back to back stem and leaf diagrams can be used to compare two sets of data.

Eg 2 The times, in seconds, taken by 10 students to complete a puzzle are shown.

9 23 17 20 12 11 24 12 10 26

Construct a stem and leaf diagram to represent this information.

2 | 0 means 20 seconds

Stem	Leaf
0	9
1	0 1 2 2 7
2	0 3 4 6

Exercise 37

1 The stem and leaf diagram shows the highest November temperature recorded in 12 European countries last year.

0 | 7 means 7°C

Stem	Leaf
0	7 9
1	0 3 4 4 4 7 8
2	0 1 2

(a) How many countries are included?
(b) What is the maximum temperature recorded?
(c) Which temperature is the mode?
(d) When the temperature in another European country is included in the data, the range increases by 2°C.
What was the temperature in that country?
Explain your answer.

2 The pie chart shows the favourite breakfast item of 36 students.

(a) What fraction of the pupils chose eggs?

(b) Work out the size of the angle for muesli.

(c) Work out how many of the 36 students chose toast.

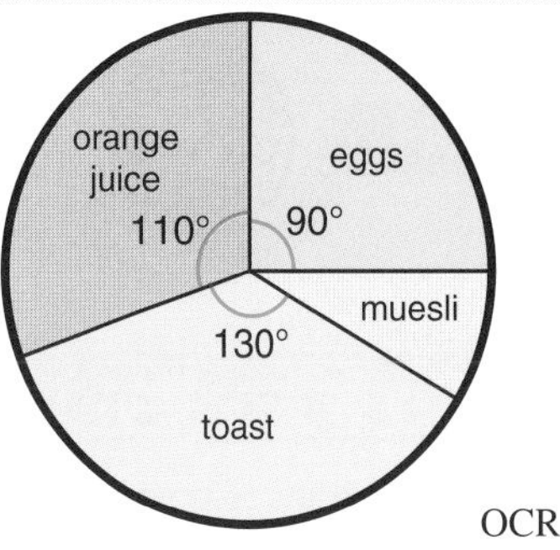

OCR

3 Pali asked 180 boys what was their favourite sport. Here are his results.

Sport	Soccer	Rugby	Cricket	Basketball	Other
Number of boys	74	25	18	37	26

(a) Draw a pie chart to show these results.

Pali also asked 90 girls about their favourite sport.
In a pie chart showing the results, the angle for Tennis was 84°.

(b) How many of these girls said that Tennis was their favourite sport?

OCR

4 The number of text messages Anila sent each day in the last two weeks is shown.

7 12 10 5 21 11 9 2 17 3 5 13 20 15

(a) Construct a stem and leaf diagram to show this information.

(b) What is the range in the number of text messages Anila sent each day?

5 A large group of 14-15 year olds was asked:

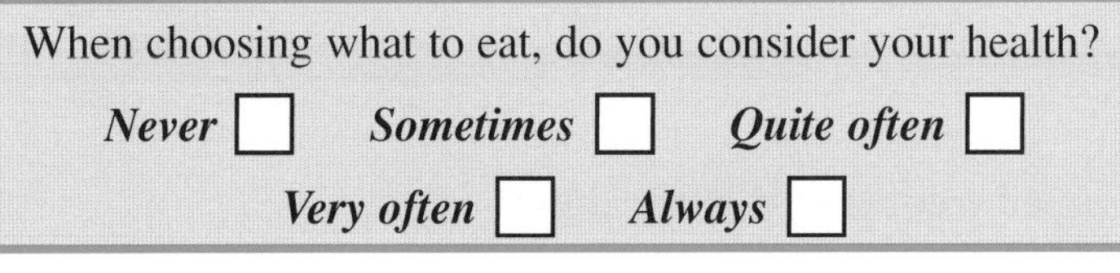
When choosing what to eat, do you consider your health?

Never ☐ *Sometimes* ☐ *Quite often* ☐ *Very often* ☐ *Always* ☐

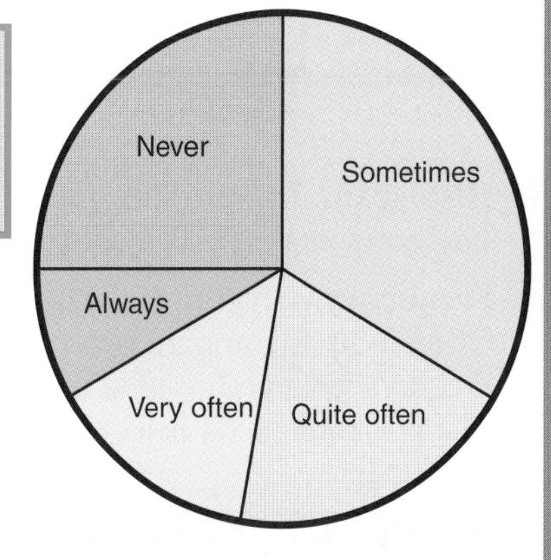

The results for the boys are shown on this pie chart.

(a) What percentage of the boys responded that they always considered their health when choosing food?

(b) What response was given by about a third of the boys?

(c) What fraction of the boys responded '*Never*'?

Here are the responses from the girls.

Response	*Never*	*Sometimes*	*Quite often*	*Very often*	*Always*
Percentage of the girls giving this response	19	45	22	9	5

(d) Show these results on a labelled pie chart.

(e) By comparing the two pie charts, write down one major difference between the girls' and boys' responses.

OCR

6 Twenty children were asked to estimate the length of a leaf.
Their estimates, in centimetres, are:

Boys									
4.5	5.0	4.0	3.5	4.0	4.5	5.0	4.5	3.5	4.5

Girls									
4.5	5.0	3.5	4.0	5.5	3.5	4.5	3.5	3.0	2.5

(a) Construct a back to back stem and leaf diagram to represent this information.

(b) Compare and comment on the estimates of these boys and girls.

SECTION 38

Time Series and Frequency Diagrams

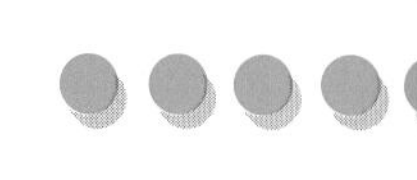

What you need to know

- A **time series** is a set of readings taken at time intervals.
- A **line graph** is used to show a time series.

Eg 1 The table shows the temperature of a patient taken every half-hour.

Time	0930	1000	1030	1100	1130	1200
Temperature °C	36.9	37.1	37.6	37.2	36.5	37.0

(a) Draw a line graph to illustrate the data.
(b) Estimate the patient's temperature at 1115.

(a)

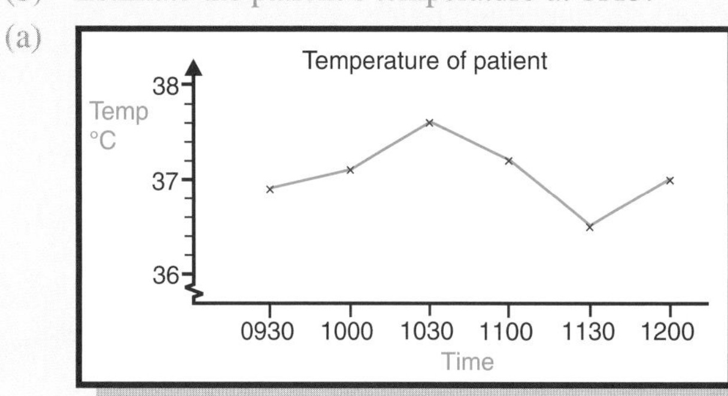

To draw a line graph:
Plot the given values.
Points are joined by lines to show the **trend**.

Only the plotted points represent **actual values**.
The lines show the **trend** and can be used to **estimate values**.

(b) 36.8°C

- **Histogram**. Used to illustrate **grouped frequency distributions.**
  The horizontal axis is a continuous scale.
- **Frequency polygon**. Used to illustrate grouped frequency distributions.
  Often used to compare two or more distributions on the same diagram.
  Frequencies are plotted at the midpoints of the class intervals and joined with straight lines.
  The horizontal axis is a continuous scale.

Eg 2 The frequency distribution of the heights of some boys is shown.

Height (h cm)	$130 \leqslant h < 140$	$140 \leqslant h < 150$	$150 \leqslant h < 160$	$160 \leqslant h < 170$	$170 \leqslant h < 180$
Frequency	1	7	12	9	3

Draw a histogram and a frequency polygon to illustrate the data.

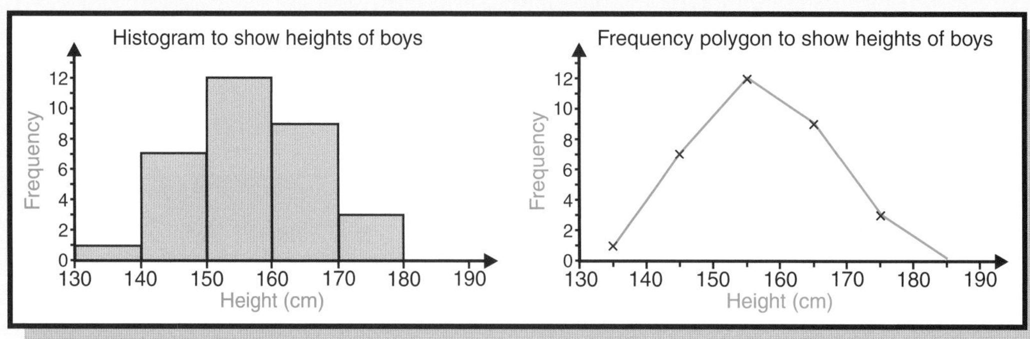

- **Misleading graphs**. Graphs may be misleading if:
  the scales are not labelled, the scales are not uniform, the frequency does not begin at zero.

Exercise 38

1 On Sunday, Alfie records the outside temperature every two hours.
The temperatures he recorded are shown in the table.

Time of day	0800	1000	1200	1400	1600	1800
Outside temperature (°C)	9	12	15	17	16	14

(a) Draw a line graph to represent the data.
(b) What is the range in the temperatures recorded?
(c) (i) Use your graph to estimate the temperature at 1300.
(ii) Explain why your answer in (c)(i) is an estimate.

2 The table shows the waiting times for patients in a doctor's surgery.

Waiting Time (t minutes)	Frequency
$0 \leqslant t < 4$	8
$4 \leqslant t < 8$	15
$8 \leqslant t < 12$	12
$12 \leqslant t < 16$	6
$16 \leqslant t < 20$	4
$20 \leqslant t < 24$	0

Draw a frequency diagram to show this information.

OCR

3 Mrs Jones collected information about lateness to school for one week.
This frequency polygon shows the information for form 10A.

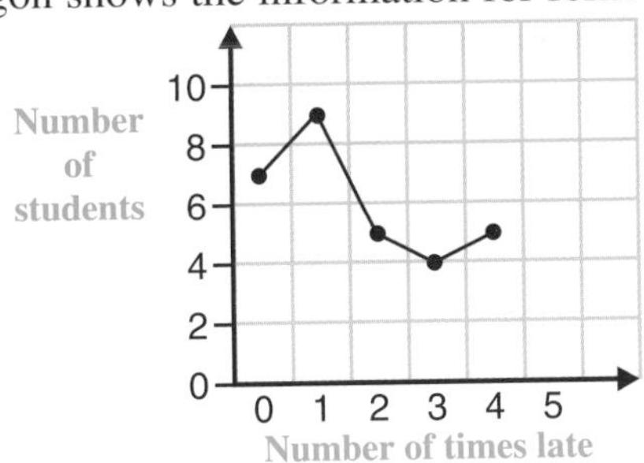

The table shows the information for form 10D.

Number of times late	0	1	2	3	4	5
Number of students	10	4	3	4	7	2

Copy the frequency polygon for form 10A.
(a) Draw the frequency polygon for form 10D on the same grid as form 10A.
(b) (i) Write down the modal number of times late for each form.
(ii) Write down the range for each form.

The mean number of times late for form 10A is 1.7.
(c) Work out the mean number of times late for form 10D.
(d) Make two comparisons between the lateness of the two forms.

OCR

4 For this diagram, give two reasons why it may be misleading.

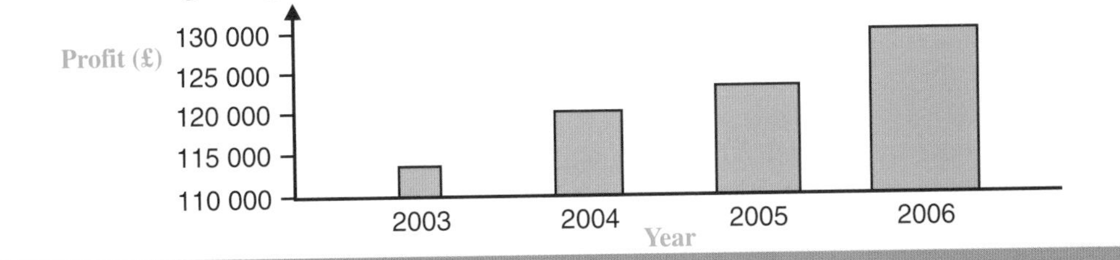

OCR

5 The table shows the times of arrival of pupils at a village primary school one day.

Time of arrival (t)	Number of pupils
$0830 \leqslant t < 0840$	14
$0840 \leqslant t < 0850$	28
$0850 \leqslant t < 0900$	34
$0900 \leqslant t < 0910$	4

(a) Draw a frequency diagram for the data.
(b) Pupils arriving after 0900 are late.
What percentage of pupils were late?

6 The graph shows the age distribution of people in a nursing home.

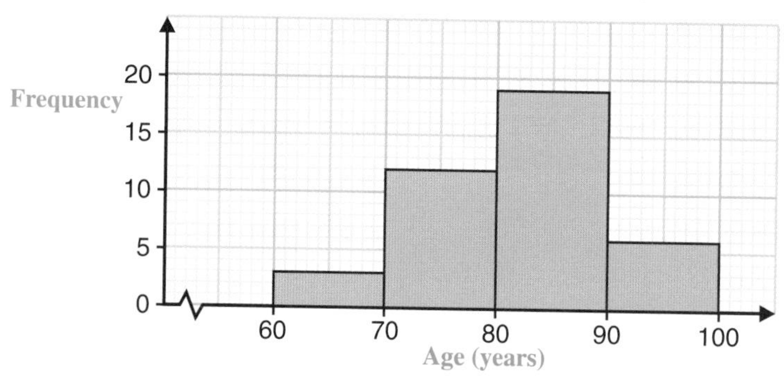

(a) Which age group is the modal class?
(b) How many people are in the nursing home?
(c) The table shows the age distribution of men in the home.

Age (a years)	$60 \leqslant a < 70$	$70 \leqslant a < 80$	$80 \leqslant a < 90$	$90 \leqslant a < 100$
Frequency	2	7	6	0

(i) Draw a frequency polygon to represent this information.
(ii) On the same diagram draw a frequency polygon to represent the age distribution of women in the home.
(iii) Compare and comment on the ages of men and women in the home.

7 The frequency polygon illustrates the age distribution of people taking part in a marathon.

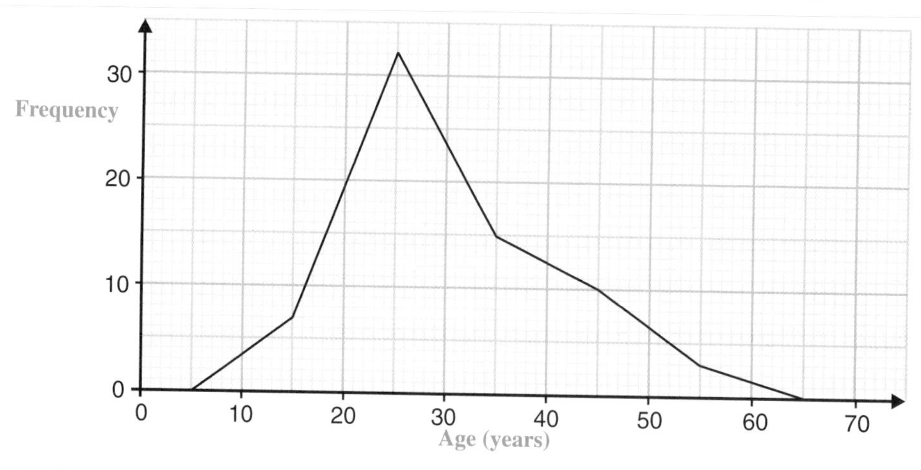

(a) How many people were under 20 years of age?
(b) How many people were over 50 years of age?
(c) How many people took part?

SECTION 39

- A **scatter graph** can be used to show the relationship between two sets of data.
- The relationship between two sets of data is referred to as **correlation**.
- You should be able to recognise **positive** and **negative** correlation. The correlation is stronger as points get closer to a straight line.
- When there is a relationship between two sets of data a **line of best fit** can be drawn on the scatter graph.
- **Perfect correlation** is when all the points lie on a straight line.

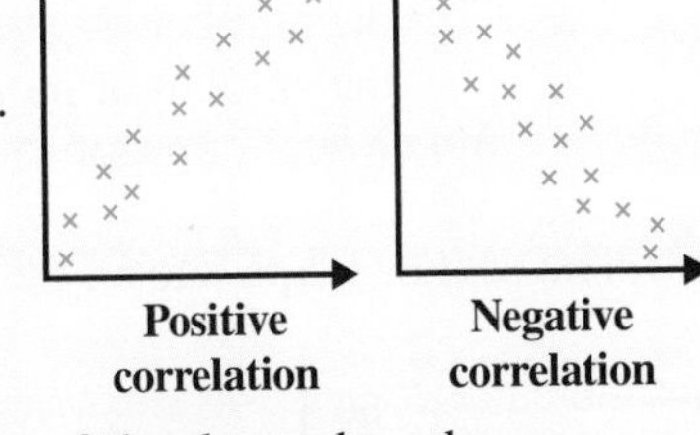

- The line of best fit can be used to **estimate** the value from one set of the data when the corresponding value of the other set is known.

Eg 1 The table shows the weights and heights of 10 girls.

Weight (kg)	33	36	37	39	40	42	45	45	48	48
Height (cm)	133	134	137	140	146	146	145	150	152	156

(a) Draw a scatter graph for the data.
(b) What type of correlation is shown?
(c) Draw a line of best fit.
(d) A girl weighs 50 kg. Estimate her height.

Mark a cross on the graph to show the weight and height of each girl.

(a)

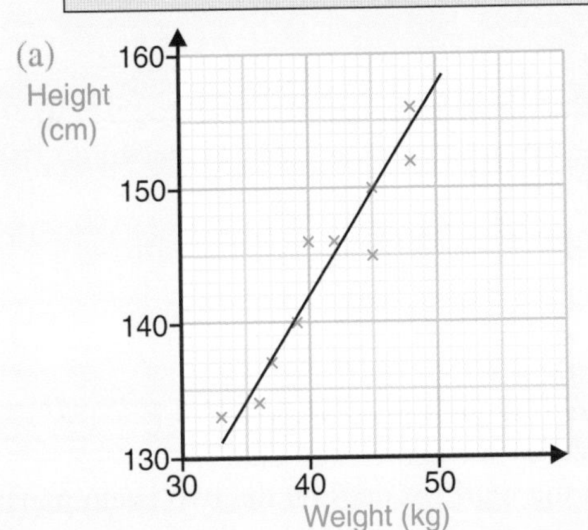

(b) Positive correlation.

(c) The line of best fit has been drawn, by eye, on the graph.

> **On a scatter graph:**
> The **slope** of the line of best fit shows the **trend** of the data.
> The line of best fit does not have to go through the origin of the graph.

(d) 158 cm.
Read estimate where 50 kg meets line of best fit.

Exercise 39

1 Ten people were asked their age and how many hours sleep they had last Friday night. The results are given in this table.

Age (years)	11	13	16	18	20	21	24	25	27	28
Hours sleep	11	11	10	9	8	7	6	7	5	6

(a) Draw a scatter graph to show this information.
(b) Choose words from this list to accurately describe the correlation shown in your scatter graph.
Weak *Strong* *Positive* *Negative* *No*
(c) Draw a line of best fit onto the diagram.
(d) Use your line of best fit to estimate the number of hours sleep of someone 15 years old.
OCR

2 The scatter graphs show the results of a survey given to people on holiday at a seaside resort.

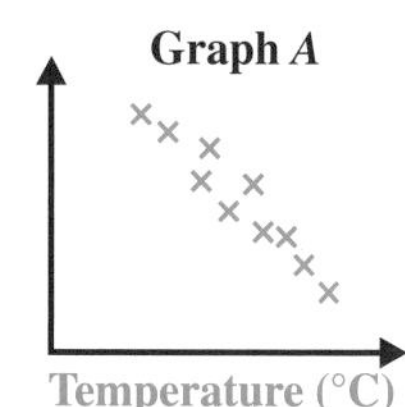

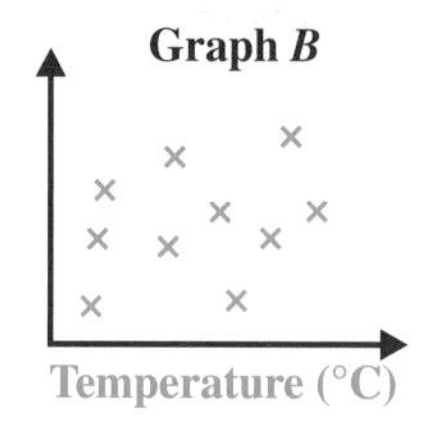

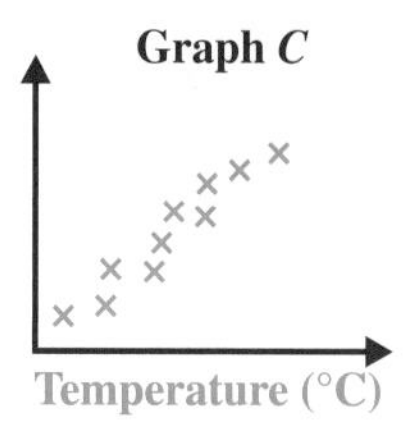

(a) Which scatter graph shows the temperature (°C) plotted against:
  (i) the number of people in the sea,
  (ii) the number of people with coats on,
  (iii) the amount of money people spend?

(b) Which scatter graph shows a positive correlation?

3 The scatter graph shows the results of candidates in two examinations in the same subject.

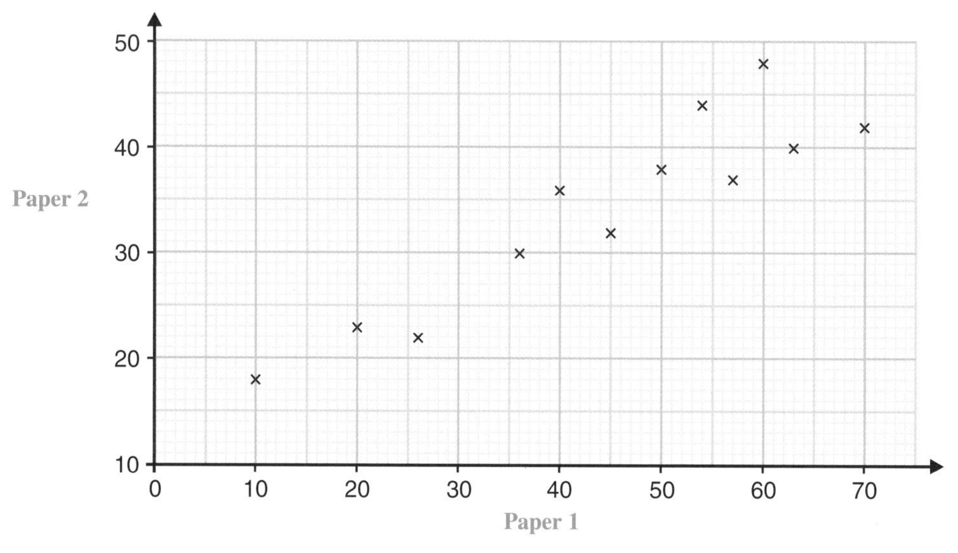

(a) One candidate scored 40 marks on Paper 1.
What mark did this candidate score on Paper 2?

(b) One candidate scored 48 marks on Paper 2.
What mark did this candidate score on Paper 1?

(c) Was the highest mark on both papers scored by the same candidate?

(d) Was the lowest mark on both papers scored by the same candidate?

(e) What type of correlation is there between the marks scored on the two exam papers?

4 A sports scientist asked eight members of a sports club how many hours per day, on average, each spent exercising.
He also measured each member's resting pulse rate.
His results are shown in the table below.

Member	A	B	C	D	E	F	G	H
Hours spent on exercise	1.5	3.5	3	2.5	2	4	1	2
Resting pulse rate (beats/minute)	72	59	62	60	70	55	70	65

(a) Draw a scatter graph to show this information.

(b) Describe the correlation between the number of hours spent on exercise and the resting pulse rate of the members.

(c) Add a line of best fit to your scatter graph.

(d) Another club member spends 6 hours per day exercising.
Explain why the line of best fit cannot be used to estimate his resting pulse rate. OCR

SECTION 40 Probability

What you need to know

- **Probability** describes how likely or unlikely it is that an event will occur.
  Probabilities can be shown on a probability scale.

Probability **must** be written as a **fraction**, a **decimal** or a **percentage**.

Less likely ← → More likely

Impossible — Certain

0 — $\frac{1}{2}$ — 1

- How to work out probabilities using **equally likely outcomes**.

$$\text{The probability of an event} = \frac{\text{Number of outcomes in the event}}{\text{Total number of possible outcomes}}$$

Eg 1 A box contains 7 red pens and 4 blue pens. A pen is taken from the box at random.
What is the probability that the pen is blue?

$$\text{P(blue)} = \frac{\text{Number of blue pens}}{\text{Total number of pens}} = \frac{4}{11}$$

P(blue) stands for the probability that the pen is blue.

- How to estimate probabilities using **relative frequency**.

$$\text{Relative frequency} = \frac{\text{Number of times the event happens in an experiment (or in a survey)}}{\text{Total number of trials in the experiment (or observations in the survey)}}$$

Eg 2 A spinner is spun 20 times. The results are shown.

4 1 3 1 4 2 2 4 3 3
4 1 4 4 3 2 2 1 3 2

What is the relative frequency of getting a 4?

$$\text{Relative frequency} = \frac{\text{Number of 4's}}{\text{Number of spins}} = \frac{6}{20} = 0.3$$

Relative frequency gives a better estimate of probability the larger the number of trials.

- How to use probabilities to **estimate** the number of times an event occurs in an **experiment** or **observation**.

Estimate = total number of trials (or observations) × probability of event

Eg 3 1000 raffle tickets are sold. Alan buys some tickets.
The probability that Alan wins first prize is $\frac{1}{50}$.
How many tickets did Alan buy? Number of tickets $= 1000 \times \frac{1}{50} = 20$

- **Mutually exclusive events** cannot occur at the same time.

When A and B are mutually exclusive events: P(A or B) = P(A) + P(B)

Eg 4 A box contains red, green, blue and yellow counters.
The table shows the probability of getting each colour.

Colour	Red	Green	Blue	Yellow
Probability	0.4	0.25	0.25	0.1

A counter is taken from the box at random.
What is the probability of getting a red or blue counter?
P(Red or Blue) = P(Red) + P(Blue) = 0.4 + 0.25 = 0.65

- The probability of an event, A, **not happening** is: P(not A) = 1 − P(A)

Eg 5 Kathy takes a sweet from a bag at random.
The probability that it is a toffee is 0.3.
What is the probability that it is **not** a toffee?

P(not toffee) = 1 − P(toffee) = 1 − 0.3 = 0.7

- How to find all the possible outcomes when two events are combined.
  By **listing** the outcomes systematically.
  By using a **possibility space diagram**.

Exercise 40

1 The list gives some words used in probability.

impossible **unlikely** **evens** **likely** **certain**

For each of these events, write down the word which describes its probability.
(a) Throwing a number less than 7 on an ordinary six-sided dice.
(b) A fair coin landing on heads.
(c) An adult chosen at random is left-handed.

2 Copy the scale. 0 ——————— 1

On your scale, draw and label arrows to show the probability of these events.
A The sun will rise tomorrow.
B The next baby born in a maternity hospital will be a girl.
C You will live to be 100.
D You will get a number greater than 1 when you roll an ordinary dice. OCR

3 Tammy invites 50 people to a party.
Each person gets a ticket with a number on it. The numbers go from 1 to 50
She is going to put all the tickets in a hat and draw one out to win a prize.
Kynan has ticket number 9.
(a) What is the probability that Kynan's ticket will win the prize?
(b) Kynan says,

**The winning ticket will be a 1-digit number or a 2-digit number.
So, there is a 50% chance of a 1-digit number winning.**

Explain why Kynan is wrong. OCR

4 Rory has two fair spinners.
The grey spinner is numbered 1, 2, 3.
The white spinner is numbered 1, 2, 3, 4.
Rory spins them both.
He adds the numbers they land on.
(a) Copy and complete the table to show all possible totals.

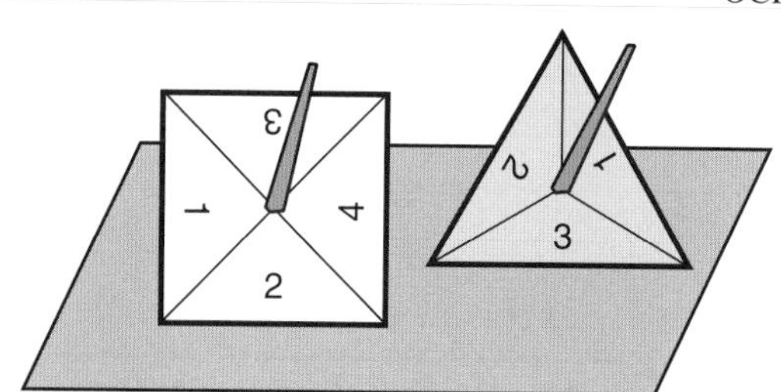

Grey spinner

		1	2	3
White spinner	**1**			
	2			5
	3			
	4			

(b) Use the table to find the probability of obtaining a total of 5. OCR

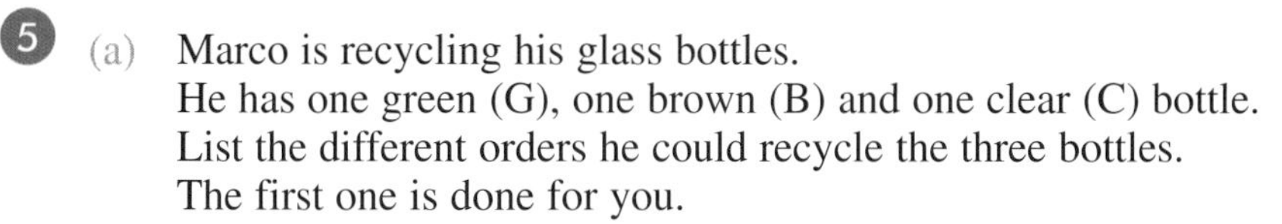

5 (a) Marco is recycling his glass bottles.
He has one green (G), one brown (B) and one clear (C) bottle.
List the different orders he could recycle the three bottles.
The first one is done for you.

G	B	C

(b) (i) Jane has 11 green, 7 brown and 2 clear bottles to recycle.
She picks the first bottle at random.
What is the probability that it is brown?

(ii) The probability that the first bottle she picks is a juice bottle is 0.4.
What is the probability that the first bottle she picks is **not** a juice bottle? OCR

6 Aimee, Georgina, Hannah and Louisa are the only runners in a race.
The probabilities of Aimee, Georgina, Hannah and Louisa winning the race are shown in the table.

Aimee	Georgina	Hannah	Louisa
0.3	0.2	0.4	

(a) Work out the probability that Louisa will win the race.
(b) Work out the probability that either Aimee or Hannah will win the race.

7 A bag contains discs of different colours.
In an experiment, Murinder took one disc out at random, noted its colour, then put it back into the bag. He repeated this 50 times.
Here are Murinder's results.

Colour	Blue	Green	Yellow	Red
Frequency	10	12	8	20

Estimate the probability that the next time Murinder takes a disc it will be red. OCR

8 Near the end of a game of Bingo, the following numbered balls still remain in the container.

2 5 8 12 17 22 23 32 34 41 42 44

(a) The next ball is drawn at random from the container.
What is the probability that the number will be less than 20?
(b) In fact, the next number drawn is 32.
Another ball is drawn at random from the container.
What is the probability that this number will be less than 20? OCR

9 Petra has 5 numbered cards. She uses the cards to do this experiment:

Shuffle the cards and then record the number on the top card.

She repeats the experiment 20 times and gets these results.

3 3 2 3 4 3 5 2 3 4
3 5 3 3 4 2 5 3 4 2

(a) What is the relative frequency of getting a 3?
(b) What numbers do you think are on the five cards?
Give a reason for your answer.
(c) She repeats the experiment 500 times.
Estimate the number of times she will get a 5.
Give a reason for your answer.

10 Jeff tosses a coin three times.
(a) List all the possible outcomes.
(b) What is the probability that he gets one head and two tails?

Handling Data Non-calculator Paper

Do not use a calculator for this exercise.

1 The pictogram shows the number of videos hired from a shop each day last week.

Day	
Monday	
Tuesday	
Wednesday	
Thursday	
Friday	
Saturday	

On Monday 6 videos were hired.

(a) How many videos does ◯◯ represent?

(b) How many videos were hired on Thursday?

70 videos were hired altogether last week.

(c) How many videos were hired on Saturday?

2 The results of a survey of the holiday destinations of people booking holidays abroad are shown in the bar chart.

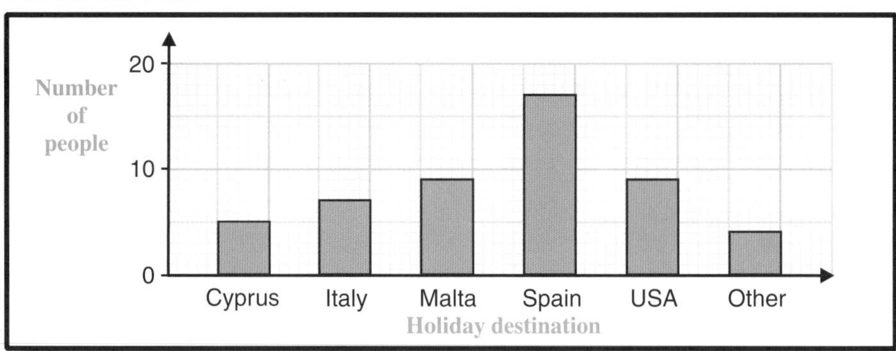

(a) Which holiday destination is the mode?

(b) How many more people are going to Spain than to Cyprus?

(c) How many people are included in the survey?

3 Alexia plays 36 holes of golf. Her score on each hole is listed.

7 5 5 8 5 4 7 3 6 6 4 4 6 5 2 6 5 5
7 4 6 6 4 5 6 4 6 5 3 8 7 5 4 7 3 8

(a) Copy and complete the frequency table below.

Score	Tally	Frequency
2		
3		

(b) Draw a bar chart to show this information.

(c) On how many holes did Alexia score 5 or less?

(d) What was the range of Alexia's scores?

OCR

4 Sarfraz records the numbers of skittles he knocks down in nine turns.

7 2 7 3 3 9 4 7 5

(a) Find the median of these numbers.
(b) Find the mode of these numbers. OCR

5 The graph shows the distribution of the best height jumped by each girl in a high jump competition.

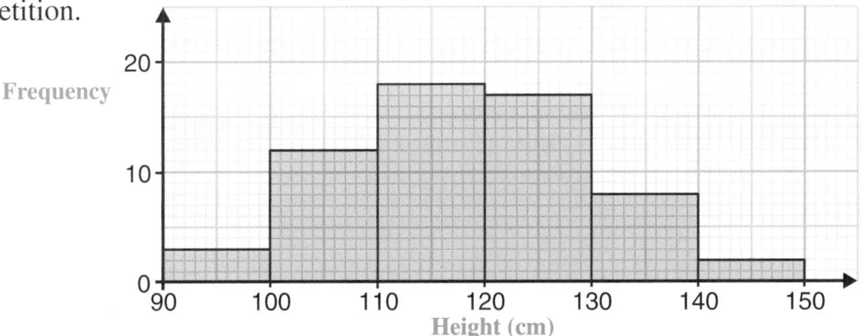

(a) How many girls jumped less than 100 cm?
(b) How many girls jumped between 100 cm and 120 cm?
(c) How many girls took part in the competition?

6 Sylvester did a survey to find the most popular pantomime.
(a) The results for children are shown in the table.

Pantomime	Aladdin	Cinderella	Jack and the Bean Stalk	Peter Pan
Number of children	45	35	25	15

(i) Draw a clearly labelled pie chart to illustrate this information.
(ii) Which pantomime is the mode?

(b) The results for adults are shown in the pie chart.
(i) 20 adults chose Aladdin.
How many adults were included in the survey?
(ii) What percentage of adults chose Cinderella?

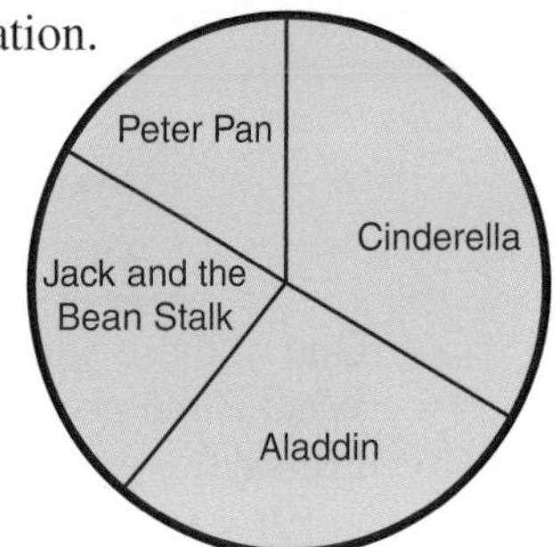

7 Two six-sided fair dice are thrown together.
The two numbers are multiplied together to give the score.
(a) Copy and complete the grid to show all possible scores.

Second dice

First dice ×	1	2	3	4	5	6
1	1	2	3	4	5	6
2	2	4	6	8	10	12
3						
4						
5						
6						

(b) What is the probability that the score is:
(i) 9, (ii) greater than 21, (iii) exactly 21? OCR

8 The lengths of 20 bolts, in centimetres, is shown.

7.4 5.8 4.5 5.0 6.5 6.6 7.0 5.4 4.8 6.4
5.4 6.2 7.2 5.5 4.8 6.5 5.0 6.0 6.5 6.8

(a) Draw a stem and leaf diagram to show this information.
(b) What is the range in the lengths of these bolts?

9 Linzi is doing a survey to find if there should be a supermarket in her neighbourhood.
This is one of her questions.

> "Do you agree that having a supermarket in the neighbourhood would make it easier for you to do your shopping and if we did have one would you use it?"

Give two reasons why this question is unsuitable in its present form.

10 Norman (N), Aled (A), Rachel (R) and Chris (C) are in the school relay team.
Norman always runs first.
The order for the other three is chosen at random.

(a) Copy and complete this table to show all the possible orders for the team.

First	Second	Third	Fourth
N	A	R	C

(b) What is the probability that Rachel will run fourth? OCR

11 The following table gives the age and blood pressure of ten men.

Age (years)	45	49	41	43	42	50	35	54	60	65
Blood pressure	124	138	125	130	138	146	117	142	154	160

(a) Draw a scatter diagram to show this information.
(b) Draw a line of best fit on your scatter diagram.
(c) Describe the correlation between age and blood pressure for this data. OCR

12 The table shows information about a group of students.

	Can speak French	Cannot speak French
Male	5	20
Female	12	38

(a) One of these students is chosen at random.
What is the probability that the student can speak French?
(b) Pru says,
"If a female student is chosen at random she is more likely to be able to speak French than if a male student is chosen at random."
Is she correct? Explain your answer.

13 A survey is carried out about the number of road accidents in a small town.

(a) The number of accidents occurring each week for 25 weeks is recorded in the frequency table below.

Number of accidents	0	1	2	3	4	5
Number of weeks (frequency)	6	3	6	7	1	2

(i) Work out the mean number of road accidents per week.
(ii) Find the range of the number of accidents.

(b) In the next 25-week period, the mean number of accidents per week was 1.4 and the range was 7.
Make two comparisons between the number of accidents per week in the two 25-week periods. OCR

14 A bag contains 50 cubes of which 7 are red.
A cube is taken from the bag at random.

(a) The probability that it is white is 0.3.
What is the probability that it is not white?
(b) What is the probability that it is either white or red?

Handling Data Calculator Paper

You may use a calculator for this exercise.

1 The table shows the number of books borrowed from a library during five days.

Day	Monday	Tuesday	Wednesday	Thursday	Friday
Number of books	40	35	30	15	50

(a) How many books were borrowed during these five days?
(b) Draw a pictogram to represent the information. Use [book symbol] to represent 10 books.

2 The temperature in the Namib Desert was measured every two hours through a 24-hour period. The results are shown in the line graph below.

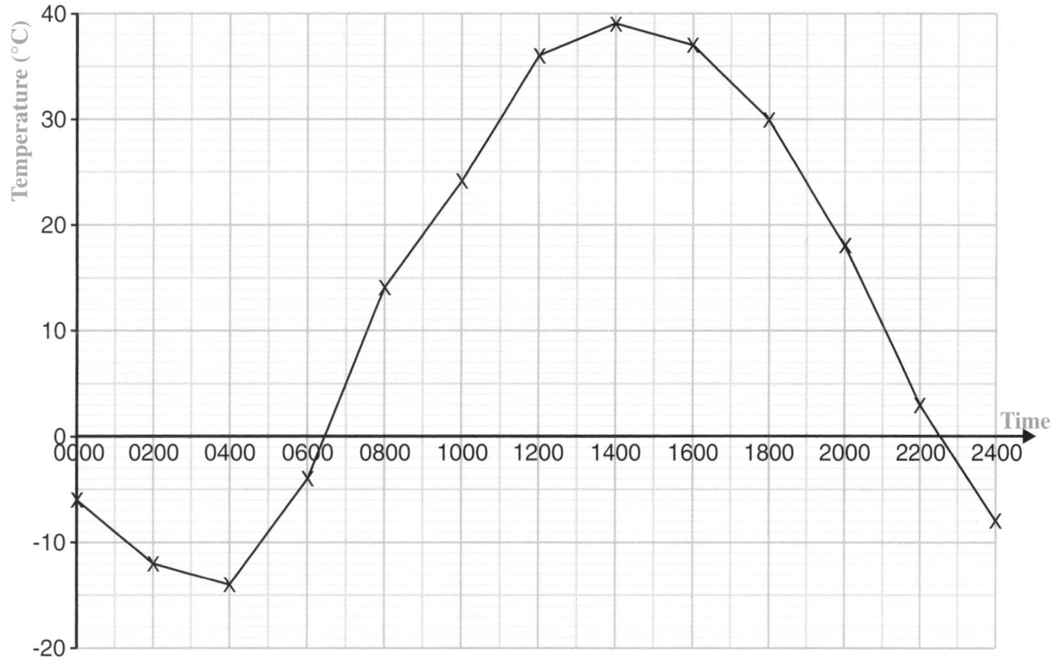

(a) What is the highest temperature recorded?
(b) What is the lowest temperature recorded?
(c) Work out the difference between the highest and the lowest recorded temperatures.
(d) Estimate the temperature at 0700 on the day that these readings were taken.
(e) Estimate for how long the temperature was above 30°C on that day. OCR

3 Pete wrote down how many litres of fuel he put in his car on his last 10 visits to the service station.
Here are his results.

20.3 30.2 26.1 14.5 35.6 27.4 16.2 38.4 26.9 18.4

Find: (a) the mean, (b) the range. OCR

4 Karina is playing a game with these cards. X

One card is taken at random from the letters.
One card is taken at random from the numbers.
(a) List all the possible outcomes.
(b) Explain why the probability of getting X 1 is not $\frac{1}{4}$.

5 The table shows the number of peas in a sample of pods.

Number of peas	1	2	3	4	5	6	7	8
Number of pods	0	0	2	3	5	7	2	1

(a) How many pods were in the sample?
(b) What is the modal number of peas in a pod?
(c) What is the range in the number of peas in a pod?
(d) Draw a bar chart to show this information.

6 The stem and leaf diagram shows the weights, in grams, of letters posted by a secretary.

1 | 5 means 15 grams

Stem	Leaf
1	5 8
2	0 4 5 6 8 8
3	1 2 3 5 7
4	2 5

(a) How many letters were posted?
(b) What is the median weight of one of these letters?
(c) What is the range in the weights of these letters?
(d) Calculate the mean weight of a letter.

7 Rovers play Wanderers at football.
The probability that Rovers win the match is 0.55.
The probability that Wanderers win the match is 0.2.
Find the probability that the result is a draw. OCR

8 A box contains 20 plastic ducks.
3 of the ducks are green, 10 are blue and the rest are yellow.
A duck is taken from the box at random.
What is the probability that it is: (a) green, (b) yellow?

9 Tom asked students in his school how they spent last Saturday morning.
The table shows the results for 200 year 11 students.

	Paid job	Doing homework	In bed	Sport	Other
Number of students	72	24	50	46	8

(a) Draw a pie chart to show Tom's results for the year 11 students.

This pie chart shows Tom's results for the year 13 students.

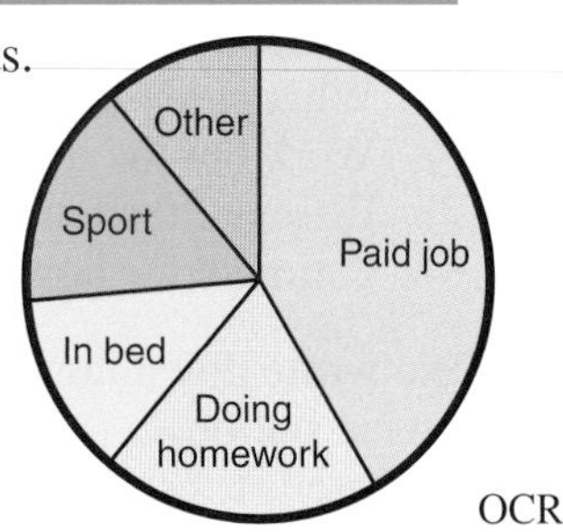

(b) (i) State one way in which the results for years 11 and 13 are similar.
(ii) State one way in which the results for years 11 and 13 are different.

OCR

10 A biased triangular spinner is divided into three sections.
The probabilities of the spinner landing on the red, blue and green sections are:

Red	Blue	Green
0.2	0.3	0.5

(a) Find the probability that the spinner will land on either the blue or the green section.
(b) The spinner is spun 250 times.
How many times would you expect it to land on red? OCR

11 The mean weight of the 16 girls in a class is 55.4 kg.
The mean weight of the 14 boys in the class is 58.2 kg.
Calculate the mean weight of the 30 pupils in the class.

12 Tariq and James collect data on the number of people in each car passing their school at lunchtime.

(a) Tariq presented the data collected on Monday in the following table.

Number of people in a car	1	2	3	4	5
Number of cars	21	17	9	2	1

(i) Find the median number of people in a car.
(ii) Calculate the mean number of people in a car.

(b) James is collecting his data on Tuesday. He says, "Based on the data I've collected so far, the probability that the next car will contain just one person is 0.25."
(i) What is the probability that the next car will contain more than one person?
(ii) James has recorded 48 cars in his survey.
How many cars contained just one person?

OCR

13 The engine size and distance travelled on one litre of petrol for each of 10 cars is summarised in the table below.

Engine size (litres)	0.6	1	1	1.1	1.6	1.8	2	2.5	2.5	3
Distance (km)	12	12	11	10	8	8	7	6	10	4

(a) Draw a scatter diagram to illustrate this information.
(b) One of these cars had been fitted with a new, efficient engine.
Identify this car by circling a point on the scatter diagram.
(c) Describe the type and the strength of the correlation shown in your diagram.
(d) (i) Draw a line of best fit on your diagram.
(ii) Estimate the distance covered by a car with a 1.3 litre engine.

OCR

14 Last Friday, a supermarket pizza stand sold the following numbers of pizzas.

Cheese and Tomato . . .	125
Ham and Pineapple . . .	96
Triple Cheese	87
Pepperoni	12

Next Friday, the supermarket expects to sell 240 pizzas in total.
Use the figures from last Friday to estimate the number of Ham and Pineapple pizzas they should make.

OCR

15 Jenni recorded the time of each of the tracks in her CD collection.
Her results are summarised below.

Time (t seconds)	Number of tracks
$120 < t \leqslant 150$	13
$150 < t \leqslant 180$	9
$180 < t \leqslant 210$	8
$210 < t \leqslant 240$	7
$240 < t \leqslant 270$	3

(a) Calculate an estimate of the mean time.
(b) Which class contains the median?
Explain how you found your answer.
(c) The random play on Jenni's CD player selects a track.
What is the probability it will last more than 240 seconds?

OCR

Non-calculator Paper

Do not use a calculator for this exercise.

1 (a) Write the number two thousand six hundred and nine in figures.
(b) Write the number 60 000 000 in words.

2 The table shows the number of cars parked in an office car park each day.

Day	Monday	Tuesday	Wednesday	Thursday	Friday
Number of cars	8	12	6	9	10

(a) How many more cars were parked on Tuesday than on Thursday?
(b) Draw a pictogram to represent the information. Use ⊕ to represent 4 cars.

3 (a) Put these numbers in order of size, smallest first.

105	30	7	19	2002

(b) Work out. (i) $105 - 30$ (ii) 19×7 (iii) $2002 \div 7$

4 Which metric unit would it be best to use for measuring
(a) the length of a pencil,
(b) the mass of one strawberry,
(c) the distance from London to Paris,
(d) the capacity of a medicine spoon? OCR

5 Write a rule for finding the next number in each sequence and use your rule to find the next number.
(a) 3, 9, 15, 21, 27, …
(b) 1, 2, 4, 8, 16, …

6 The diagram shows a rectangle and a triangle drawn on 1 cm squared paper.
(a) How many lines of symmetry has
(i) the rectangle,
(ii) the triangle?
(b) What is the perimeter of the rectangle?
(c) What is the area of the triangle?

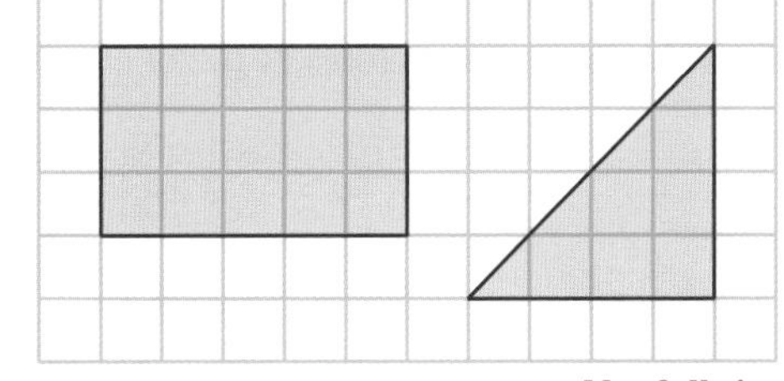

Not full size

7 (a) A roll costs d pence. How much will 5 rolls cost?
(b) A cake costs 25 pence more than a roll. How much does a cake cost?

8 (a) Work out. (i) $710 - 481$
(ii) 13×100
(iii) $132 \div 1000$
(b) Write down (i) two factors of 21,
(ii) two multiples of 21,
(iii) a square number between 20 and 30. OCR

9 (a) 6 people go to watch a pop concert. It cost them £32 each.
Work out the total cost.
(b) 599 people travel to the pop concert by coach. Each coach could carry up to 52 people.
Work out the smallest number of coaches needed. OCR

10 Pino is making three wooden shelves. The shelves will be 1.20 m, 55 cm and 90 cm long.
The wood can be bought in 2 m, 3 m or 4 m lengths.
(a) Which one of the lengths should he buy so that he has the least amount of wood left over?
Show how you decide.
(b) What length of wood is left over? Give your answer in centimetres. OCR

11 Use the formula $P = 5m + 2n$ to find the value of P when $m = 4$ and $n = 3$.

12 (a) What fraction of the rectangle is shaded?
Give your answer in its simplest form.
(b) What percentage of the rectangle is **not** shaded?

13 Sports Wear Ltd. hire out ski-suits. The cost of hiring a ski-suit is calculated using this rule.

Four pounds per day plus a fixed charge of five pounds.

(a) How much would it cost to hire a ski-suit for 8 days?
(b) Heather paid £65 to hire a ski-suit.
For how many days did she hire it?

14 (a) On graph paper plot the points $P(4, 1)$ and $Q(2, -5)$.
(b) Find the coordinates of the midpoint of the line segment PQ.

15 (a) Copy the following shapes and draw in all the lines of symmetry.
If there are none, write '*None*'.

(i) 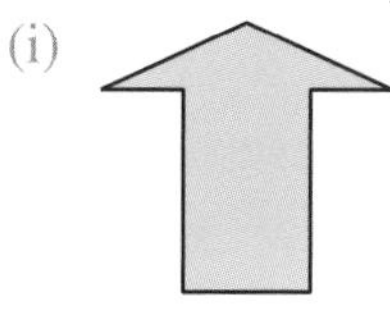
(ii) 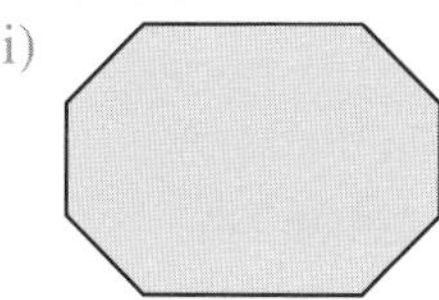
(iii) 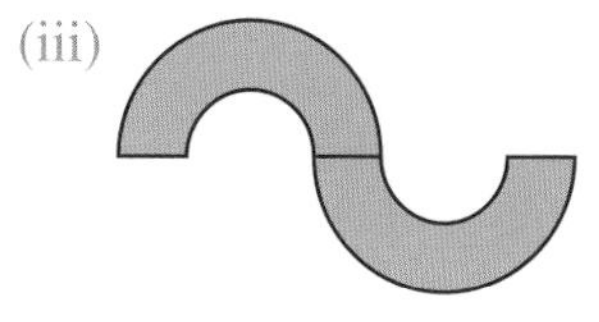

(b) 
What is the order of rotational symmetry of this shape?

(c) Copy this diagram and draw two squares on it so that it has rotational symmetry of order 3.

OCR

16 The prices, in pence, of 9 different loaves of bread are shown.

23, 36, 69, 49, 38, 55, 82, 69, 29

Work out (a) the range in prices, (b) the median price, (c) the mean price.

17 The temperature at noon was 5°C. At midnight the temperature was 7°C colder.
What was the temperature at midnight?

18 In the diagram, XY is a straight line.
Find the size of the angle a.

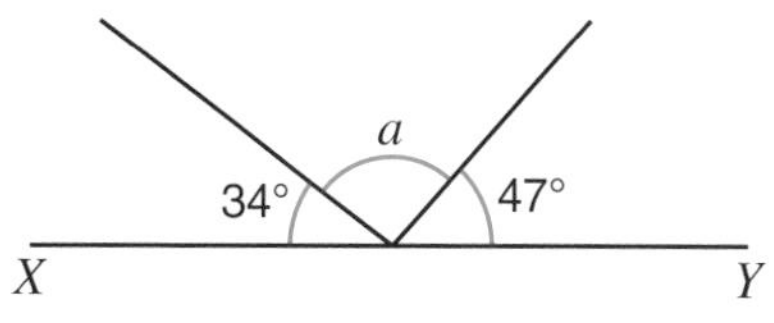

19 (a) Simplify $3p + 7q + p - 5q + 5p$.
(b) Solve. (i) $\frac{w}{2} = 1.5$ (ii) $2z - 3 = 4$

OCR

20 (a) Draw and label the lines $y = x + 1$ and $x + y = 3$ for values of x from -1 to 3.
(b) Write down the coordinates of the point where the lines cross.

21 In the diagram, angle $BCD = 76°$, $AC = BC$ and ACD is a straight line.
Work out the size of angle BAC.

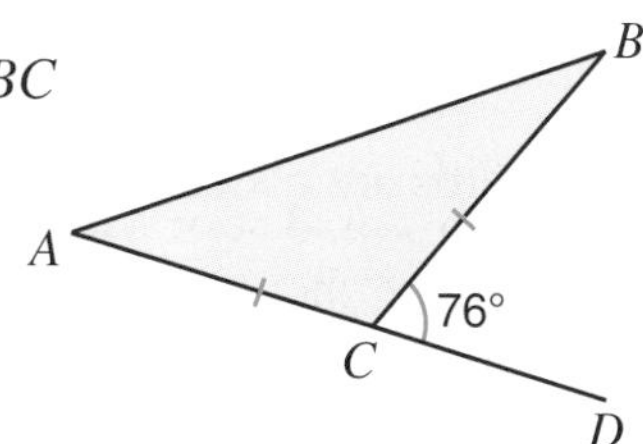

22 (a) Leopold visits a café.
List all the different 2-course meals he could have.

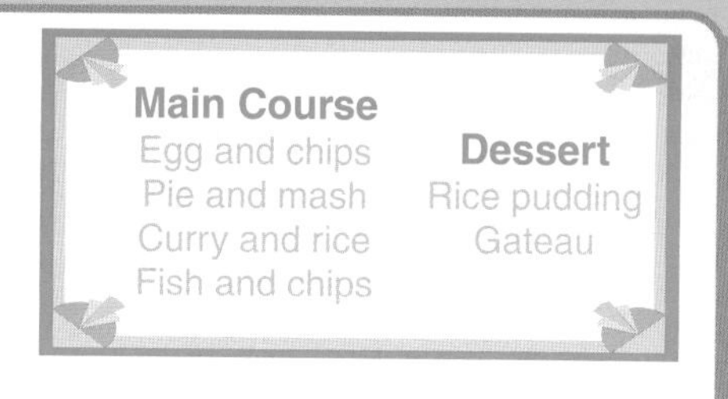

(b) One lunchtime the café sells 90 main courses.
The table shows how many of each dish are sold.

Dish	Egg	Pie	Curry	Fish
Frequency	10	19	20	41

Draw and label a pie chart to show this information.

OCR

23 To calculate the number of mince pies, m, to make for a Christmas Party for p people, Donna uses the formula $m = 2p + 10$.

(a) How many mince pies would she make for a party of 12 people?

(b) Donna makes 60 mince pies for another party.
How many people are expected at this party?

24 Maggie has a box of chocolates. It contains milk, plain and white chocolates.
Maggie chooses a chocolate at random.
The probability of choosing a milk chocolate is $\frac{3}{8}$.

(a) There are 40 chocolates in the box.
How many are milk chocolate?

(b) The probability of choosing a plain chocolate is $\frac{1}{2}$.
What is the probability of choosing a white chocolate?

OCR

25 Here is part of the timetable for the Swanage railway.

Station	Time
Norden (N)	10:30
Corfe Castle (C)	10:32
Harman's Cross (H)	10:41
Swanage (S)	10:53

Here are the distances between the stations.

N —1 mile— C —2 miles— H —3 miles— S

(a) Draw a distance-time graph for this journey.

(b) Mark with an arrow the section of the graph that shows when the train was travelling fastest.

(c) Explain one way in which your graph does not show exactly how the train is really travelling.

OCR

26 A cuboid has a volume of $90\,\text{cm}^3$.
The base of the cuboid measures 3 cm by 6 cm.
Calculate the height of the cuboid.

27 Glen buys this car for £3600.
How much must he pay each month?

OCR

28 (a) Work out. $\frac{3}{5} + \frac{1}{4}$

(b) Beth and Lucy share £80 in the ratio 3 : 1.
Work out how much each of them receives.

OCR

29 The numbers on these cards are coded. The sum of the numbers on these 3 cards is 41.

x	$2x - 1$	$3x$

(a) Form an equation in x.

(b) By solving your equation, find the numbers on the cards.

30 (a) In the diagram the lines AB and CD are parallel.
They are crossed by two straight lines.
Find angle x, giving a reason for your answer.

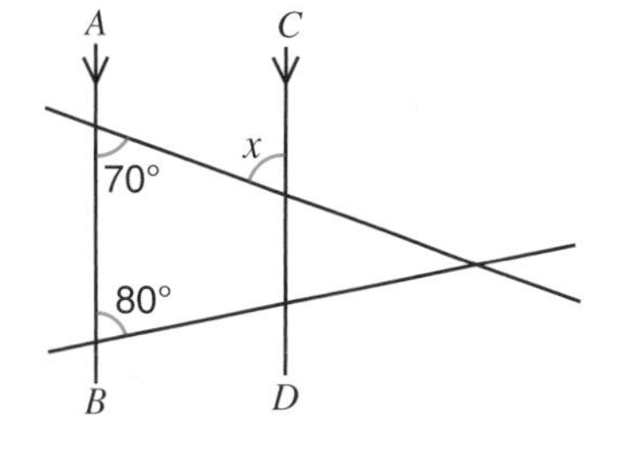

(b)

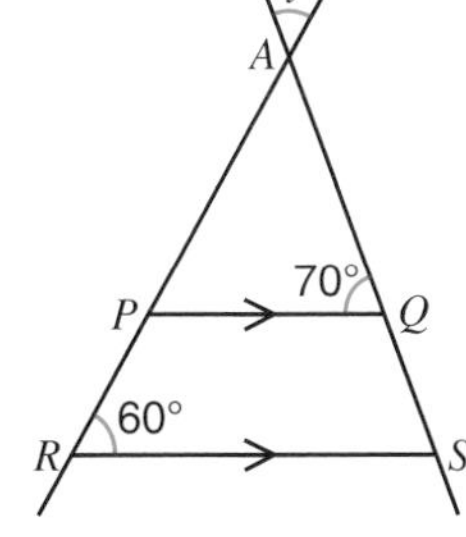

In this diagram, the lines PQ and RS are parallel.
Find angle y, showing how you obtained your answer.

OCR

31 Write 72 as a product of its prime factors.

32 The diagram shows the positions of shapes P, Q and R.

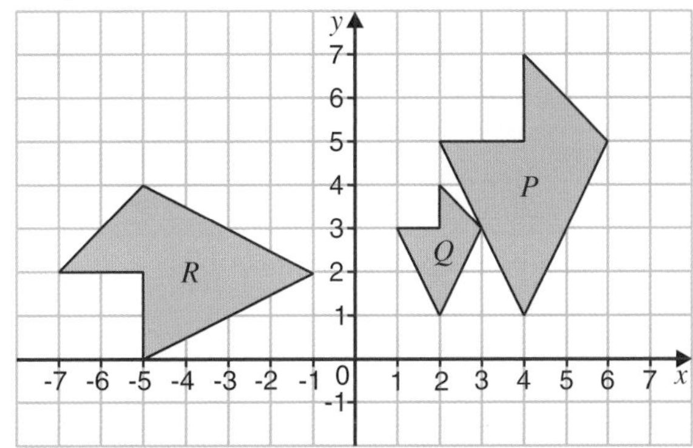

(a) Describe fully the single transformation which takes P onto Q.
(b) Describe fully the single transformation which takes P onto R.

33 (a) Jack's foot length is 20 cm. His height is 1.6 m.
Write, in the form $1 : n$, the ratio: Jack's foot length : Jack's height
(b) Janet's foot length and height are in the same ratio as Jack's.
Her foot length is 14 cm. Work out her height.
(c) Jagdeep's foot length is 21 cm, correct to the nearest centimetre.
Write down his greatest and least possible foot length.

OCR

34 (a) Copy and complete the table of values for $y = x^2 - 3x + 1$.

x	-1	0	1	2	3	4
y		1	-1			5

(b) Draw the graph of $y = x^2 - 3x + 1$ for values of x from -1 to 4.
(c) Use your graph to find the value of y when $x = 1.5$.
(d) Use your graph to solve the equation $x^2 - 3x + 1 = 0$.

35 Cocoa is sold in cylindrical tins. The height of a tin is 7.9 cm. The radius of a tin is 4.1 cm.
Use approximations to estimate the volume of a tin. Show all your working.

36 (a) Factorise. $6x + 15$
(b) Solve this equation. $3(x - 7) = x - 4$

OCR

37 (a) Make t the subject of the formula. $W = 5t + 3$
(b) Simplify. $m^2 \times m^3$

38 These formulae represent quantities connected with containers, where $\boldsymbol{a}$, $\boldsymbol{b}$ and $\boldsymbol{c}$ are dimensions.

$$\mathbf{2(ab + bc + cd)} \qquad \mathbf{abc} \qquad \sqrt{\mathbf{a^2 + b^2}} \qquad \mathbf{4(a + b + c)}$$

Which of these formulae represent lengths? Explain how you know.

Calculator Paper

You may use a calculator for this exercise.

1 (a) Write these numbers in order, smallest first:

5, −7, 0, 3.5, 10

(b) What is the value of the 8 in the number 18 326?

2 (a) Look at this pattern of dots.

Pattern 1 Pattern 2 Pattern 3 Pattern 4

(i) Copy and complete the table.

Pattern	1	2	3	4	5
Number of dots	4	8			

(ii) How many dots would be needed for Pattern 100? Explain how you worked out your answer.

(b) Here are the first six terms of a sequence.

31 30 28 25 21 16 …

Write down the next term.
Explain how you worked out the answer.

(c) Look at this pattern.

$8^2 - (2 \times 8) + 1 = 49$

$7^2 - (2 \times 7) + 1 = 36$

Complete the next row.

$6^2 - (2 \times \ldots) + 1 = \ldots$

OCR

3 Copy the diagram and draw a reflection of the shape in the mirror line PQ.

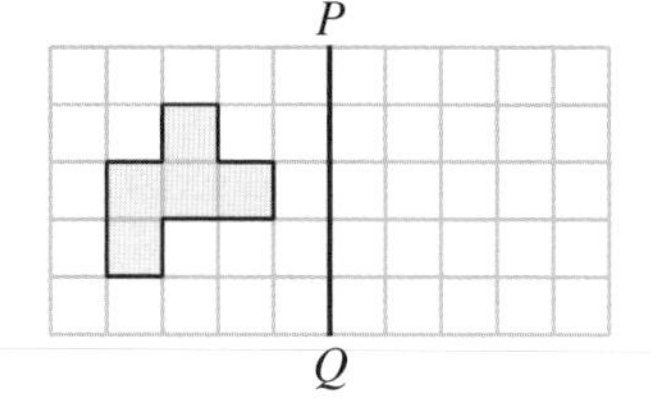

4 These words are used in probability.

impossible unlikely evens likely certain

A fair, six-sided dice is thrown once.
Which word describes the probability of getting each of these events?

(a) An odd number.

(b) A number **more** than 5.

5 British Airways has four daily flights from London to Lisbon.

Depart London	0935	1220	1405	1825
Arrive Lisbon	1210	1510	1645	2100

(a) How long does the 0935 flight take? Give your answer in hours and minutes.

(b) How much longer does the 1220 flight take than the 1405 flight?

(c) Anneka has an appointment in the centre of Lisbon at 5 pm. The centre of Lisbon is one hour from the airport. Which is the latest plane she should catch?

OCR

6. The diagram shows two gear wheels.
The large wheel has 24 teeth.
The small wheel has 12 teeth.
Describe what happens to the small wheel when the large wheel is turned through 90° in a clockwise direction.

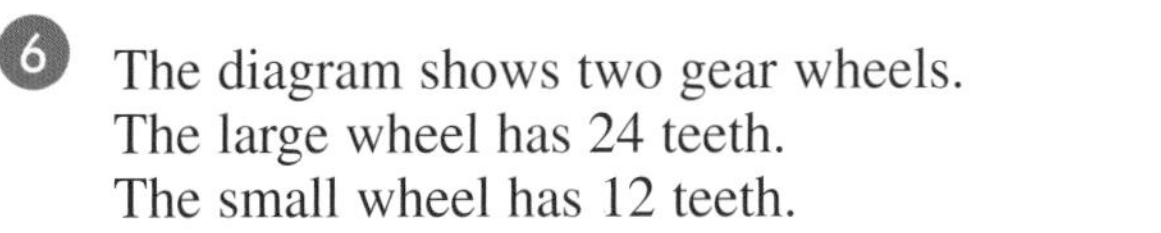

7. In France a bicycle costs 120 euros. £1 = 1.40 euros.
How much is the bicycle in £s, correct to the nearest £?

8. Which of these triangles are congruent?

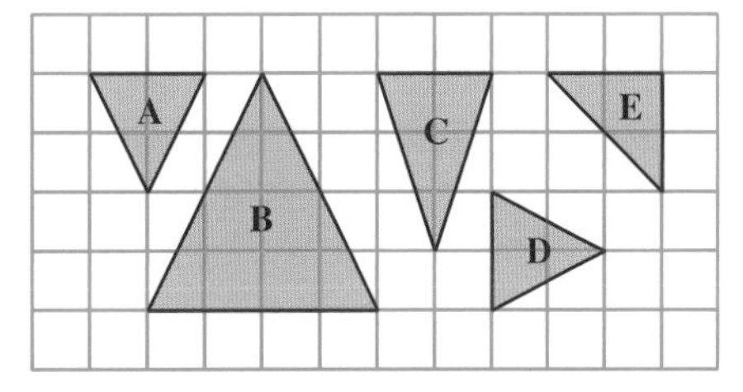

9. Find the missing numbers in each of these:
(a) $50 - 18 = \square$ (b) $25 + \square = 42$ (c) $\square \div 4 = 6$

10. This pie chart shows which end-of-term activity a class of pupils preferred.

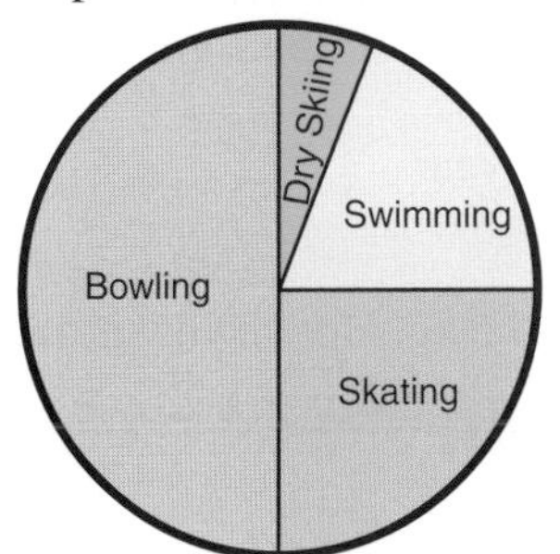

(a) Half of the pupils prefer Bowling.
What percentage is this?
(b) What percentage prefer Skating?
(c) There are 32 pupils in the class.
How many pupils prefer Skating?

OCR

11. Copy the diagram and draw two more shapes so that the final pattern has rotational symmetry of order 4.

12. Gloria makes hats.
This formula is used to calculate her weekly pay, in pounds.

Pay = 3 × Number of hats + 75

(a) Last week, Gloria made 48 hats. How much was she paid?
(b) This week, Gloria was paid £285. How many hats did she make?

13. (a) Simplify $5g - 3g + 2g$.
(b) Solve the equations (i) $3n = 12$, (ii) $3m + 1 = 10$.
(c) Find the value of $2h^2$ when $h = 3$.
(d) Find the value of $3p + q$ when $p = -2$ and $q = 5$.

14. Write $\frac{1}{3}$, 0.5, 40% in order, smallest first.

15. The diagram shows a cuboid.
(a) Draw an accurate net of the cuboid.
(b) Work out the area of the net.

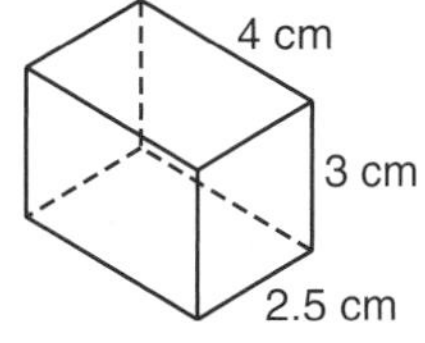

16. Cheri is paid a basic rate of £6.40 per hour for a 35-hour week.
Overtime is paid at $1\frac{1}{2}$ times the basic rate.
Last week she worked 41 hours. Calculate her pay for last week.

17 Here is an Input-Output diagram.

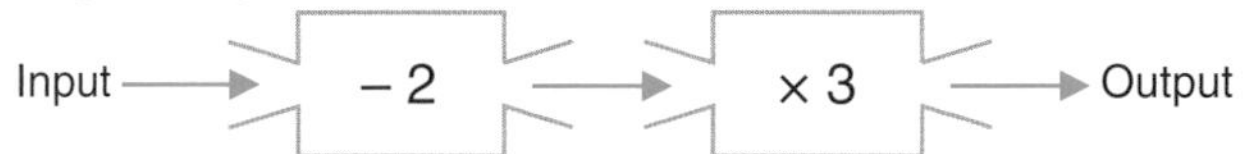

(a) What is the Output when the Input is −1?
(b) What is the Input when the Output is 9?

18 (a) Calculate. (i) 12.1^2 (ii) $\sqrt{46}$ (iii) $\frac{5}{0.5 \times 0.4}$
(b) Copy and complete this sentence.
One million is the same as ten to the power of … OCR

19 A fairground ride is decorated with 240 coloured lights.
(a) 15% of the lights are red. How many red lights are there?
(b) 30 of the 240 lights are not working. What percentage of lights are not working?

20 Jo buys a new car.
(a) She fills it with 36 litres of petrol.
Roughly, how many gallons is 36 litres?
(b) The distance from home to work is 5 miles.
She works 3 days a week.
Roughly, how many kilometres will she drive to and from work each week?
(c) One day, she visits her sister. She travels at an average speed of 45 mph for 4 hours.
How many miles does she travel?
(d) The value of her car is £5400. Its value falls by 20% in one year.
By how much does the value of her car fall? OCR

21 Jacob is 3.7 kg heavier than Isaac. The sum of their weights is 44.5 kg. How heavy is Jacob?

22 Bob cycles from home to work. The travel graph shows his journey.

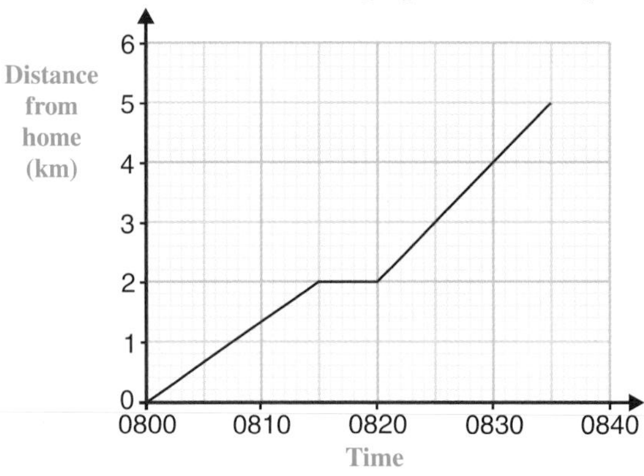

(a) On his way to work Bob stopped to buy a newspaper.
At what time did he stop?
(b) (i) During which part of his journey did Bob cycle fastest?
Give a reason for your answer.
(ii) Calculate his average speed in kilometres per hour for this part of his journey.

23 A recipe for a fruit crumble includes these ingredients.

200 g flour	125 g margarine	100 g sugar	20 g ginger	750 g rhubarb

(a) Paul has 300 g of flour. He uses it all to make a larger fruit crumble with the recipe.
What weight of sugar should he use?
(b) Sally has 600 g of rhubarb. She uses it to make a smaller crumble with the recipe.
What weight of margarine should she use? OCR

24 A bill for a meal for six people was £128.40. The bill included £34.68 for drinks.
What percentage of the bill was for drinks? OCR

25 (a) A regular polygon has 9 sides.
Work out the sum of its interior angles.

(b) In the diagram, PQ is parallel to RS.
Find the size of angles a, b and c.

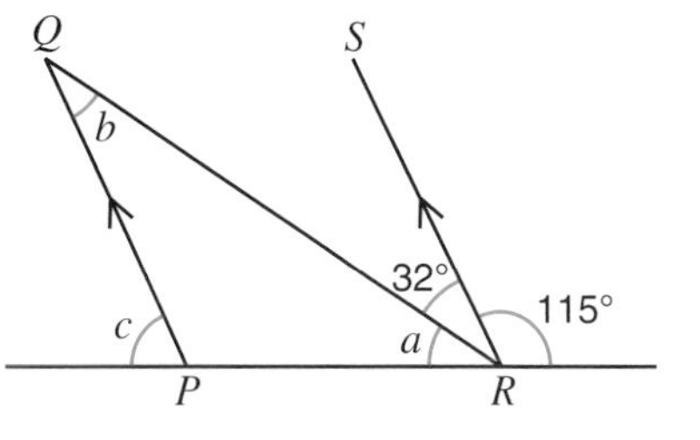

26 (a) Multiply out and simplify where possible. (i) $x(x - 3)$ (ii) $5(2y + 1) + 3y$

(b) Factorise. (i) $6a + 9$ (ii) $2b^2 + b$

(c) Solve. (i) $\frac{x}{5} = 20$ (ii) $4x + 3 = 2x - 6$

OCR

27 A hang glider flies 2.8 km on a bearing of 070° from P to Q and then 2 km on a bearing of 200° from Q to R.

(a) Make a scale drawing to show the flight of the hang glider from P to Q to R.
Use a scale of 1 cm to 200 m.

(b) From R the hang glider flies directly back to P.
Use your drawing to find the distance and bearing of P from R.

28 Some students took part in a sponsored silence.
The frequency diagram shows the distribution of their times.

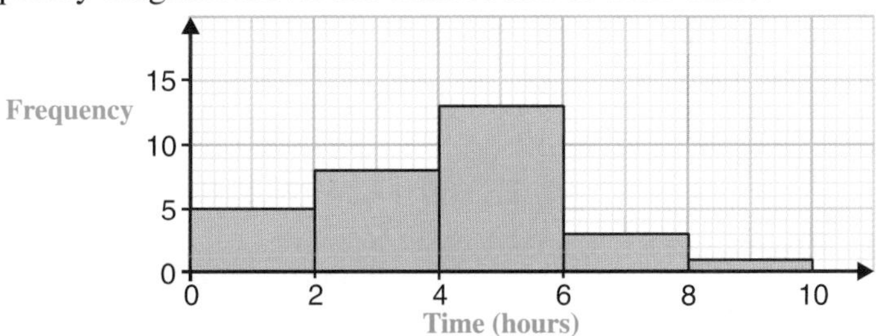

(a) How many students took part?

(b) Which time interval contains the median of their times?

(c) Calculate an estimate of the mean of their times.

29 The diagram shows a solid triangular prism made of metal.
The triangle has base 4.2 cm and height 3.5 cm.

(a) Find the area of the triangle.

(b) The length of the prism is 9.8 cm.
Find the volume of the prism.

(c) The mass of the prism is 500 g.
What is the density of the metal?

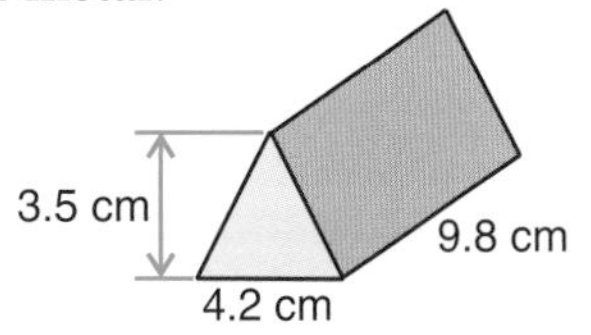

OCR

30 Use a trial and improvement method to find a solution to the equation $x^3 + x = 57$.
Show all your working and give your answer correct to one decimal place.

31 The diagram shows a semi-circle with diameter AB.
C is a point on the circumference.
$\angle ACB = 90°$. $AC = 6$ cm and $CB = 8$ cm.
Calculate the area of the shaded triangle as a percentage of the area of the semi-circle.

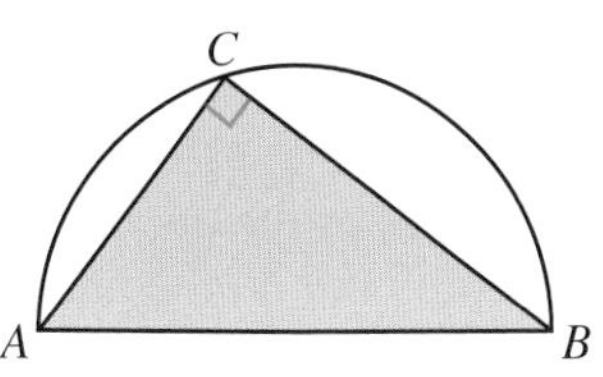

32 A biased, six-sided dice has a probability of $\frac{3}{8}$ of landing on a 6.
The dice is thrown 500 times.
About how many times would you expect it **not** to land on 6?

OCR

33 (a) Write down the values of n, where n is an integer, which satisfies the inequality

$$-1 < n + 2 \leqslant 3.$$

(b) Solve the inequality $2x + 3 < 4$.

Answers

SECTION 1

Exercise 1 Page 1

1. (a) 111 994
   (b) Fourteen thousand and seventy-six
2. (a) 3 thousands
   (b) 117, 100, 85, 23, 9
3. (a) 3, 57, 65, 71 (b) 50, 65
   (c) 50 (d) 3 and 57
4. (a) 81
   (b) (i) 35 (ii) 100 (iii) 10
   (c) (i) 1005 (ii) 191 (iii) 183
5. (a) 466 km (b) Jean's journey by 34 km
6. (a) 108 (b) 90
7. (a) (i) 78 (ii) 89 (iii) 23
   (b) **85 + 73 + 62 = 220**
8. (a) (i) 2358 (ii) 8523 (b) 6165
9. (a) $55555 \times 7 = 388885$
   (b) The number of 8's should be one less than the number of 5's.
10. (a) 295
    (b) 11 676
11. £181 per month
12. £3072
13. 7
14. 7
15. 50 boxes
16. (a) 18 (b) 12
    (c) 3
17. 70 kg
18. 115 cm
19. 125 girls
20. £6894
21. 218 days
22. (a)  $1 + 2 + 3 + 4 = \frac{4 \times 5}{2} = 10$
    (b) 5050

SECTION 2

Exercise 2 Page 3

1. 0.065 and 0.9
2. 1.08, 1.118, 1.18, 1.80
3. (a) £22.15 (b) 674
   (c) 9 (d) 16.2
4. 4.9 kg
5. (a) £3.80 (b) 5
6. £137.70
7. £27.45
8. 38 pence
9. £229.12
10. There are two figures after the decimal points in the question but only one in the answer.
11. (a) 0.16 (b) 7.8
12. (a) 7.36 (b) 3.2 (c) 230
13. (a) $\frac{3}{10}$ (b) $\frac{3}{100}$ (c) $\frac{33}{100}$
14. 40 minutes
15. $0 < m < 1$. E.g. $\frac{1}{2}$, $\frac{2}{5}$
16. 11.5 p/kg. Bag: 38 p/kg, sack: 26.5 p/kg.
17. £2.50 per kilogram
18. 17.76792453

SECTION 3

Exercise 3 Page 6

1. (a) 626 (b) 630 (c) 600
2. (a) 480 (b) 4000
3. 19 500
4. 300 km + 100 km = 400 km
5. (a) 9000, 1300, 1700
   (b) $\frac{9000}{1300 + 1700} = \frac{9000}{3000} = 3$
6. $\frac{3000 \times 40}{100} = £1200$
7. (a) $40 \times 20 = 800$
   (b) (i) $2000 \div 40$ (ii) 50
8. (a) 2.8 (b) $80 \div 30$
9. $\frac{£20\,000}{50} = £400$
10. (a) 100 is bigger than 97, **and** 50 is bigger than 49.
    (b) Smaller. 1000 is smaller than 1067, **and** 50 is bigger than 48.
11. $60\,000 \div 60 = 1000$
12. (a) £307.20
    (b) $20\,000 \times £10 = £200\,000$
13. 49.5 m
14. (a) 635 (b) 644
15. (a) $40 \times £7$
    (b) Bigger. 40 is bigger than 39, and £7 is bigger than £6.95.
16. 17 boxes
17. $\frac{400 + 200}{40} = \frac{600}{40} = 15$. Answer is wrong.
18. (a) $£60\,000 \div 20 = £3000$
    (b) $\frac{50 \times 40}{0.5} = \frac{2000}{0.5} = 4000$
19. (a) $\frac{20 \times 60}{100} = \frac{1200}{100} = 12$
    (b) $12 - 10.875 = 1.125$
20. (a) $\frac{9000}{10} \times 90\text{p} = £810$
    (b) 9000 is larger than 8873, 10 is smaller than 11, and 90 is larger than 89.9.
21. Minimum: £95, maximum: £104.99
22. (a) 14.95 (b) 15.0
23. (a) 680 (b) 700
24. No. For example, an answer of 0.01634… is 0.02 to 2 d.p. and 0.016 to 2 s.f. 0.016 is more accurate.
25. (a) 8.299 492 386 (b) 8.30
26. 6.4
27. 0.3

SECTION 4

Exercise 4 — Page 8

1. (a) −3°C (b) −13°C
2. (a) Oslo (b) Warsaw
3. 140 m
4. −9, −3, 0, 5, 7, 17
5. (a) (i) Venice (ii) Oslo (b) −6°C
6. £57 overdrawn (−£57)
7. (a) 6 degrees (b) Between 1200 and 1800
8. (a) 4 (b) 11 (c) −3
9. (a) 6 (b) 5 (c) −3 (d) −3
10. 7 degrees
11. (a) 3454 m (b) 2°C
12. (a) 5 (b) −30 (c) −2 (d) −5
13. 20°F
14. (a) −20 (b) −15
15. 5

SECTION 5

Exercise 5 — Page 10

1. (a) $\frac{2}{5}$ (b) Shade 4 parts.
2. £24
3. (a) 7 (b) 0.75 (c) $\frac{2}{6}$ and $\frac{10}{30}$
4. (a) £30 (b) $\frac{5}{16}$ (c) $\frac{5}{12}, \frac{11}{24}, \frac{1}{2}, \frac{5}{8}, \frac{2}{3}$
5. (a) $\frac{7}{10}$ as $\frac{4}{5} = \frac{8}{10}$ (b) E.g. $\frac{5}{12}$
6. (a) $\frac{7}{9}$ (b) 0.375 (c) $\frac{7}{8}$
7. 8 jars
8. $\frac{2}{3}$
9. (a) $\frac{2}{3}$ (b) $\frac{3}{14}$
10. £46
11. £3.14
12. (a) $\frac{1}{10}$ (b) $\frac{3}{20}$
13. (a) 0.167 (b) 1.7, 1.67, 1.66, $1\frac{1}{6}$, 1.067
14. $\frac{12}{25}$
15. $\frac{9}{20}$
16. (a) $1\frac{5}{12}$ (b) $\frac{9}{10}$
17. £3.80 per kilogram.
18. (a) 0.1875 (b) $\frac{6}{11}$

SECTION 6

Exercise 6 — Page 13

1. (a) 4 (b) 18 (c) 7
2. (a) 1, 2, 3, 6, 9, 18 (b) 35 (c) 9 has more than 2 factors: 1, 3, 9 (d) 1, 2, 3, 6
3. (a) 16 (b) 12, 24 (c) 27, 64
4. (a) 10 (b) 2
5. (a) 49 (b) 8
6. No. $2^2 + 3^2 = 4 + 9 = 13$
   $(2 + 3)^2 = 5^2 = 25$
7. (a) 27 (b) 16 (c) 0.09 (d) 80
8. (a) 55 (b) 8100 (c) 200
9. (a) (i) 1000 (ii) 2 (iii) 0.36 (b) (i) 27 (ii) 23 or 29
10. (a) 72 (b) 17 (c) 112
11. (a) 7^{-1} (or $\frac{1}{7}$) (b) 13 (c) $2^3 \times 3 \times 5$
12. 4
13. 30 seconds
14. (a) 5 (b) 0.25 (c) $\sqrt{225}$. $\sqrt{225} = 15$, $2^4 = 16$
15. (a) $x = 9$ (b) $x = 3$
16. (a) 3^5 (b) 5^3 (c) 2^2
17. (a) 8 and 9 (b) 8.37
18. (a) 0.14 (b) 175.616
19. 7.86
20. 6.76
21. (a) 5.45 (b) 2.937
22. 3.465
23. (a) 10.657 (b) $\sqrt{\frac{4}{0.2^2}} = \sqrt{\frac{4}{0.04}} = \sqrt{100} = 10$

SECTION 7

Exercise 7 — Page 15

1. (a) 30% (b) 25% (c) 40%
2. 0.02, 20%, $\frac{1}{2}$
3. (a) 2 pence (b) 15 kg (c) £45
4. Daisy.
   Daisy scored $\frac{4}{5} = 80\%$.
5. £13.50
6. 1
7. 700 people
8. £1.20
9. 5%
10. (a) 126 (b) 30%
11. £72
12. £8.45 per hour
13. Rosie.
    $\frac{52}{75} = 69.3\%$
14. 91 152 people
15. 25%
16. 36%
17. £203
18. 46%
19. 81
20. (a) £192 (b) 36%

SECTION 8

Exercise 8 — Page 17

1. £213.75
2. (a) 1455 (b) (i) 35 minutes (ii) £10.40
3. Amart. Amart: $4 \times 45p = £1.80$
   Bazda: $5 \times 47p - 50p = £1.85$
4. 12 days
5. £62
6. £5.20
7. Small bar.
   Large: 2.66 g/p
   Small: 2.78 g/p
8. £480
9. (a) 6.24 euros per kilogram (b) £2.69
10. £46.75

SECTION 9

Exercise 9 — Page 19

1. £16.25
2. £1876.25
3. £614
4. £40.74
5. £9 × 30 = £270
6. (a) £1894 (b) £189.40
7. (a) £10.80 (b) £87.30
8. £5.60

9. Costs: $50 \times 3 \times £10 + 60 \times £20 = £2700$
Income: $30 \times £90 + 10 \times £20 = £2900$
So, he has enough money to pay bills.
10. (a) £7.40 (b) £155.40
11. (a) 45 pence (b) 34 minutes
12. £267.38
13. (a) 0030 or 30 minutes past midnight.
(b) £11.70
14. £1725.90
15. £88.80
16. £49.94

SECTION 10

Exercise 10 — Page 21

1. 7 : 3
2. 3 cm by 4 cm
3. 8 large bricks
4. 75%
5. 7.5 kg
6. £25
7. (a) 100 ml
(b) 200 ml
8. 4 : 1
9. 840 males
10. £87
11. (a) 70%
(b) 9 women
12. (a) 150 g
(b) 100 ml
13. £1.96
14. £517.50
15. (a) £24
(b) Emma: £54
Rebecca: £36
16. 1 : 20 000
17. £930

SECTION 11

Exercise 11 — Page 23

1. 64 km/h
2. $1\frac{1}{2}$ hours
3. 165 km
4. 40 minutes
5. 2.5 km
6. 2 km
7. 3 km/h
8. (a) 40 mph
(b) 1116
9. (a) 60 mph
(b) 190 miles
10. 5 hours 42 minutes
11. 47 mph
12. Yes.
$\frac{65}{80} \times 60 = 48.75$ mins
Arrives 1029
13. 28.8 mph
14. (a) $1\frac{1}{2}$ hours
(b) 50 mph
15. 10 m/s
16. 9 g/cm^3
17. 19 g
18. 259.3 people/km^2

Number

Non-calculator Paper — Page 25

1. (a) (i) 6, 10, 16, 61, 100 (ii) 193
(b) (i) 63 (ii) 2000 (iii) 25
2. (a) Five thousand and thirty-one (b) 2604
(c) (i) 5830 (ii) 6000
(d) (i) 541 (ii) 209
3. (a) Missing entries are:
£2.40, 80p, 25p. Total £3.95
(b) £6.45
4. (a) 6 and 14 or 9 and 11
(b) Any two: 9, 11, 15, 27
(c) 15 (d) 27 (e) 11
5. 7 hours 47 minutes
6. (a) (i) 75 (ii) 133 (iii) 286
(b) 121 (c) (i) 2 (ii) 10
7. (a) 0.035, 0.462, 0.5, 0.5089, 0.54
(b) $\frac{3}{100}$ (c) 52% (d) 0.6
(e) $\frac{7}{10}$ $\left(\frac{7}{10} = 70\%\right)$
8. (a) 74 pence (b) £4.90
9. £4752
10. **9845**
11. (a) 1, 2, 3, 4, 6, 8, 12, 24
(b) 1, 2, 3, 4, 6, 12
(c) 6
12. (a) −32°C (b) 42 degrees (c) 15°C
13. 1500 m
14. £30
15. 12 footballs
16. (a) £1.50 (b) £10.91
17. (a) 3628.3 (b) 4000
18. £266.70
19. (a) (i) 100 000 (ii) 68 (iii) 72 (iv) 0.9
(b) 5^4, $5^4 = 625$, $4^5 = 1024$
(c) 50
20. (a) $\frac{7}{10}$ (b) £23.40 (c) (i) 0.08 (ii) 80
21. (a) $\frac{1}{2}, \frac{3}{5}, \frac{5}{8}, \frac{2}{3}, \frac{3}{4}$
(b) $\frac{9}{40}$
(c) (i) $\frac{13}{20}$ (ii) $\frac{1}{6}$ (iii) $\frac{8}{15}$
(d) $4\frac{4}{5}$ or 4.8
22. (a) 8100 (b) 36 000
23. (a) 34.7 (b) $50 \times 300 = 15\,000$
24. 180 cars
25. (a) £27 (b) $\frac{7}{15}$
26. £70
27. (a) £42.75 (b) £34.20
28. $50 \times £10 + 100 \times £7 + 30 \times £15$
$= £500 + £700 + £450 = £1650$
He does not have enough money.
29. (a) $\frac{3}{8}$ (b) 25%
30. (a) 16 km/h (b) 1106
31. Smallest: 245 straws, largest: 254 straws
32. (a) (i) 586 740 (ii) 0.462
(b) $\frac{3000}{50 \times 20} = \frac{3000}{1000} = 3$
33. (a) (i) 4 (ii) 10 (b) $4\frac{5}{12}$
34. (a) 3^7 (b) 3^5 (c) 3^2
35. (a) 200 (b) $2 \times 3 \times 5$
36. (a) Blackberries. Same reduction in price from a smaller amount.
(b) 120 grams (c) 20% (d) 8.5 kg
37. 10
38. (a) (i) $2^4 \times 3$ (ii) $2^2 \times 3^3$ (b) 432

Number

Calculator Paper

Page 28

1. (a) 5 (b) 3570 (c) Four tenths
2. (a) −7, −1, 0, 5, 13 (b) 20
3. (a) 27 minutes
   (b) (i) 1250 (ii) 1 hour 42 minutes
   (c) £10.90
4. £1.17
5. (a) 7, £5.50 change (b) 6 packets
6. (a) 10, 6 (b) 864
7. (a) 15 (b) 8 000 000
8. (a) 0.78 (b) 0.3, $\frac{8}{25}$, 33%, $\frac{1}{3}$
9. 3
10. (a) 83 pence (b) 92 pence
11. (a) $4 \times 5 \times 30 = 600$ (b) 10 times too big
12. (a) 150 (b) 55.9%
13. (a) 290 euros (b) £27.02
14. £363.60
15. 24.1 kg
16. 20
17. (a) (i) 52.037 (ii) 52.04 (iii) 52.0
    (b) (i) 6.7 (ii) 8.41
    (iii) 65 (iv) $3.\dot{3}$
18. (a) 474 units (b) £55.95
19. (a) 200 g (b) 30 (c) 13 pence
20. (a) 4 : 3 (b) 60.7% (c) £2.10
21. Small.
    Small: $\frac{180}{36} = 5$ g/p Large: $\frac{300}{63} = 4.76$ g/p
22. 0.5 litres
23. 68 mph
24. £115.15
25. (a) 16.5 (b) 0.08, 11%, $\frac{9}{20}$, 0.7
    (c) £2.10 per kg
26. 12.5%
27. £144
28. (a) (i) 290 (ii) $\frac{600 \times 30}{80 - 20} = \frac{18\,000}{60} = 300$
    (b) 4
29. (a) £352 (b) 1150 litres
30. (a) 0.625 (b) 63.125
31. (a) £6324 (b) 8.4%
32. 3150

SECTION 12

Exercise 12

Page 31

1. £$9k$
2. $(t + 5)$ years
3. (a) $2a$
   (b) $5c + 2d$
4. $(3x + 2y)$ pence
5. (a) $6m$ (b) $m + 2$
   (c) m^3
6. $7x + y$
7. $(4x + 200)$ degrees
8. $(5d + 15)$ pence
9. (a) $p + 6t + 16$
   (b) $6p - 10t$
10. (a) $7E$
    (b) $9s$
11. $a + a$ and $2a$ $2(a + 1)$ and $2a + 2$
    $2a + 1$ and $a + a + 1$ a^2 and $a \times a$
12. (a) (i) $3x + 3$ (ii) $x + 2y$
    (b) (i) $2x + 6$ (ii) $x^2 - x$
    (c) (i) $2x - 5$ (ii) $13 + 3x$
    (d) (i) $2(a - 3)$ (ii) $x(x + 2)$
13. (a) $10a^2$ (b) $6gh$ (c) $2k$ (d) 3
14. (a) £xy (b) £$y(x - 5)$
15. $(500 - 6x)$ pence
16. (a) $3ab - 2a - b$ (b) $8x + 19$
17. (a) $6g$ (b) h^8 (c) m^3
18. (a) $2a + 6b$ (b) $a + 4$
19. $4x - 1$
20. (a) y^5 (b) x^3 (c) z^2
21. $x^2 + 2x - 15$
22. (a) a^4 (b) b^4 (c) c^3 (d) d^4
23. (a) $2(2x + 3)$
    (b) (i) $6y - 9$ (ii) $x^3 - 2x^2$ (iii) $a^2 + ab$
    (c) $x^2 - x$
24. $m^2 - 5m + 6$

SECTION 13

Exercise 13

Page 33

1. (a) 15 (b) 9 (c) 5 (d) 15
2. (a) $a = 5$ (b) $b = 2$ (c) $c = 4$
3. 9
4.

Input	3	5	−2
Output	9	13	−1

5. (a) 5 (b) 4
6. (a) $x = 5$ (b) $x = 23$ (c) $x = 7.4$
7. (a) $x = 10$ (b) $x = 5$
   (c) $x = 4$ (d) $x = -6$
8. (a) $x = -1$ (b) $x = \frac{1}{2}$
   (c) $x = 5\frac{1}{2}$ (d) $x = -0.8$

SECTION 14

Exercise 14

Page 34

1. (a) $x = 4$ (b) $x = -1$
   (c) $x = 5$ (d) $x = 21$
2. (a) 30 (b) 0 (c) 9
3. $x = 7$, $y = 5$, $z = -3$
4. (a) $x = 16$ (b) $x = \frac{1}{2}$
5. (a) $x = 11$ (b) $x = 6$
6. (a) $x = 4$ (b) $x = 6$ (c) $x = 9$
7. (a) $x = 8$ (b) $x = 1$
   (c) $x = -4$ (d) $x = 2.5$
8. (a) $x = 1\frac{1}{3}$ (b) $x = -2$
9. (a) $x = -1.5$ (b) $x = 2.5$
   (c) $x = 0.6$ (d) $x = 1.5$
10. $x = 3$

11. (a) $n + (n + 3) + (2n - 1) = 4n + 2$
(b) $4n + 2 = 30, \quad n = 7$

12. $n + (2n + 5) = 47, \quad 3n + 5 = 47, \quad n = 14$.
Larger box has 33 chocolates.

13. (a) $3x + 90 = 285$ (b) 65 pence

14. $x = -2$

15. (a) $x = 7$ (b) $x = 0.6$

16. $x = 5$

17. (a) $(n - 7)$ pence
(b) $10n + 5(n - 7) = 445, \quad n = 32$
Party hat costs 25 pence.

18. (a) $x^2 - 3x$ (b) $y = -2$

19. (a) $q = -2$ (b) $t = 1\frac{1}{2}$

20. (a) When $x = 2, \quad x^3 - 5x - 8 = -10$
When $x = 3, \quad x^3 - 5x - 8 = 4$
(b) $x = 2.8$

21. $x = 3.7$

SECTION 15

Exercise 15

Page 36

1. 4

2. £37

3. (a) 2 (b) −8 (c) 8 (d) −15

4. $H = -13$

5. (a) 27 (b) 36

6. (a) £70 (b) £104 (c) 16 miles

7. (a) (i) 170 (ii) −16
(iii) No. If n is even, answer is odd.
(b) $b = 6$

8. $L = -10$

9. $A = -11$

10. (a) $F = 19$ (b) $F = 22$

11. (a) 18 (b) 32

12. 24

13. $T = 100$

14. Reuben calculated:
$(2 \times -3)^2 + 5 = (-6)^2 + 5 = 36 + 5 = 41$
He should have calculated:
$2 \times (-3)^2 + 5 = 2 \times 9 + 5 = 18 + 5 = 23$

15. $C = 200 - 35n$

16. (a) 40 km (b) $K = \frac{8M}{5}$ (c) $M = 37.5$

17. (a) $A = 10$ (b) $n = \frac{c + 5}{10}$

18. $m = \frac{n - 3}{p}$

19. $n = 14$

20. $r = \frac{4C}{3}$

21. (a) $v = -7$ (b) $a = \frac{v - u}{t}$

SECTION 16

Exercise 16

Page 38

1. (a) 21, 25 (b) 30, 26

2. (a) 17 (b) 81 (c) $\frac{1}{16}$

3. 37, 60

4. (a)

(b) Missing entries are:
10, 12, 14, 16
13, 16, 19, 22
(c) (i) $2 + (2 \times \text{pattern number})$
(ii) $1 + (3 \times \text{pattern number})$
(d) (i) 37 (ii) Pattern 4

5. (a) 14
(b) No. Number must be (multiple of 3) − 1.

6. (a) Multiply the last term by 3. (b) 405

7. (a) 10, 28
(b) (i) 2 (ii) Halve the previous term to find the next term, etc.
(c) 46

8. (a) Pattern 20 has 58 squares.
$3 \times (\text{pattern number}) - 2$
(b) $3n - 2$

9. (a) 3, −3 (b) 0 (c) $4n + 1$

10. (a) $2n + 3$ (b) $5n - 4$

11. (a) $88 - 8 = 80, \quad 80 - 4 = 76$
(b) $2n - 3$ (c) −3, 0, 5

SECTION 17

Exercise 17

Page 41

1. (a) $R(-6, 2), \quad S(3, -4)$
(b) $T(-3, 0), \quad U(0, -2)$

2. (a)

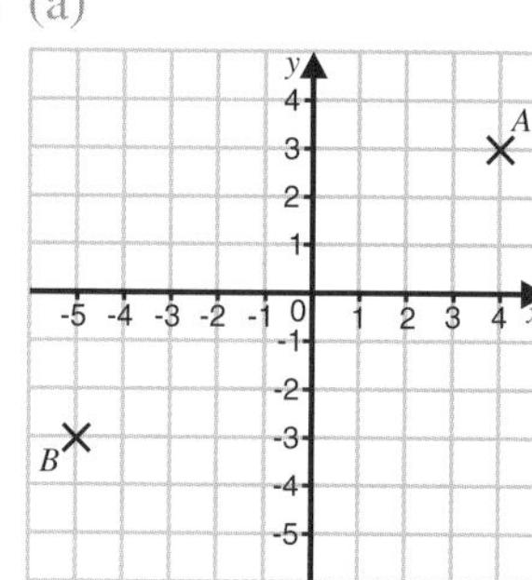

$p = -2$

3. (a)

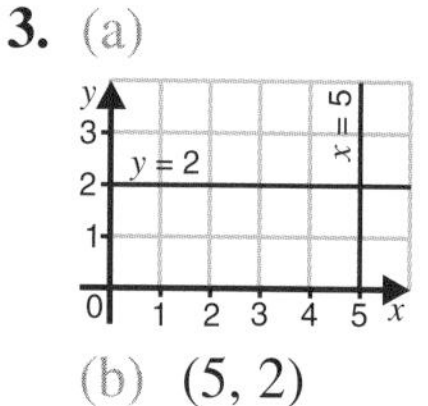

(b) (5, 2)

4. (a) Missing entries are: −2, 4

(b)

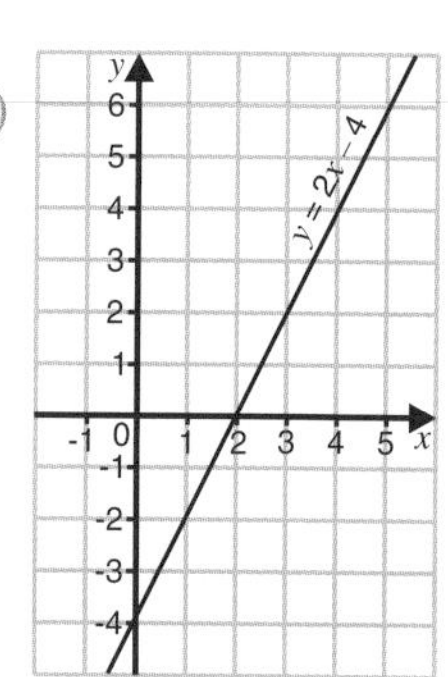

5. (a)

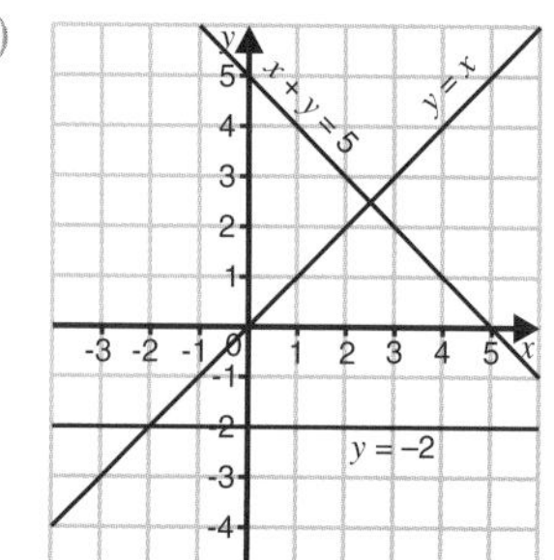

(b) $x + y = 5$

6. Straight line joining (0, -2) and (5, 13).

7. (a)

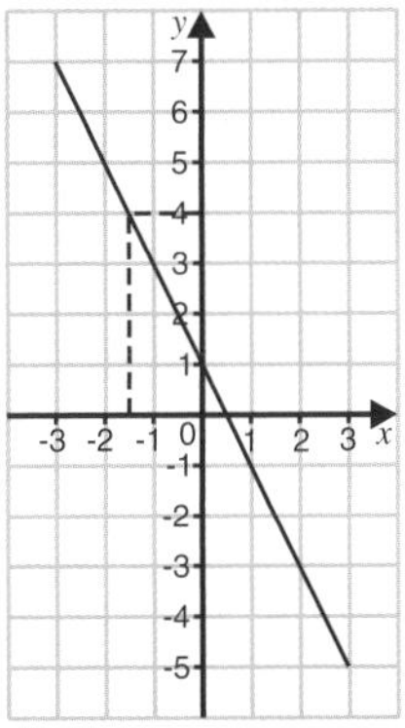

(b) $y = 4$

8. (a) $P(0, 3)$, $Q(6, 0)$ (b) $m = 5.5$

9. (a) $y = 3$ (b) $y = \frac{1}{2}x + 1$

10. (a) Missing entries are: -6, 3

(b)

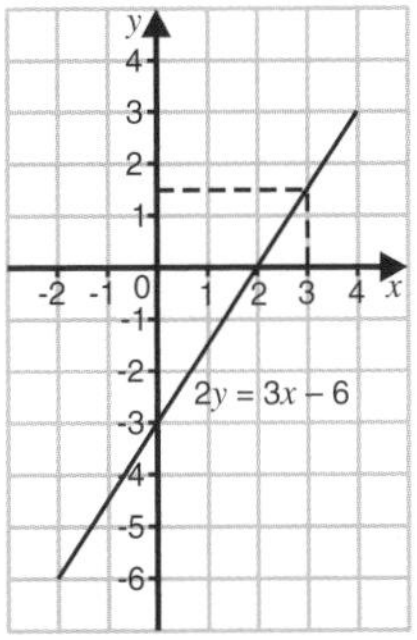

(c) $x = 3$

11. (a)

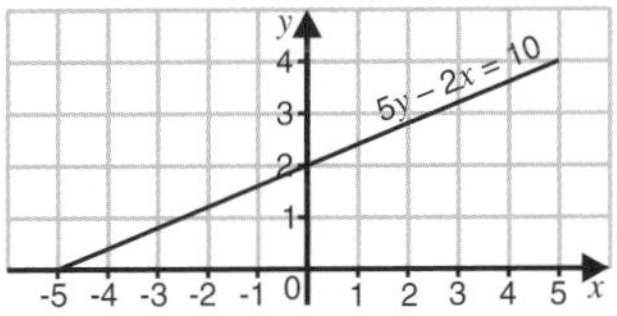

(b) $y = 1.2$

12. (a)

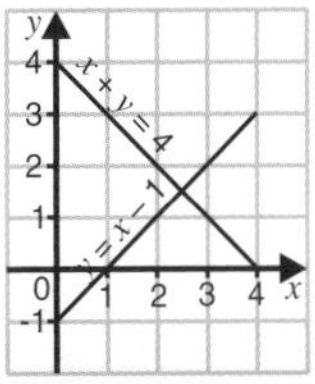

(b) (2.5, 1.5)

SECTION 18

Exercise 18

Page 42

1. (a) 25 miles (b) 16 km

(c) From graph, 10 miles = 16 km.
So, 500 miles = 50×16 = 800 km.

2. (a) 8°C (b) 30 minutes

(c) 30 minutes

3. (a)

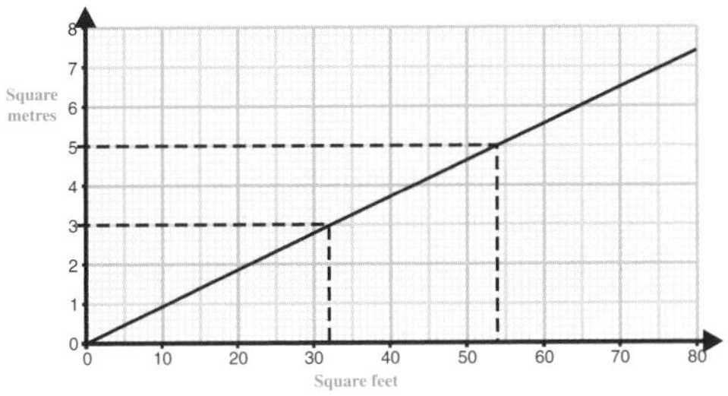

(b) (i) 54 square feet (ii) 3 square metres

4. (a)

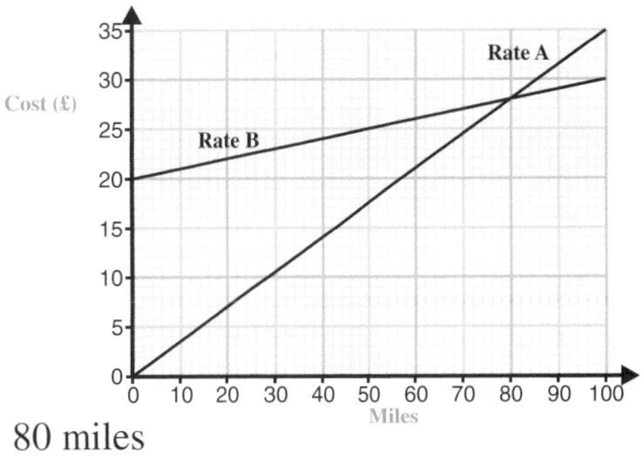

(b) 80 miles

5. (a) Between 1045 and 1130 (b) 60 km/h

6. (a) £480 (b) £100 (c) £980

7. (a) 4 minutes (b) 300 m

(c) Returning home

8. (a)

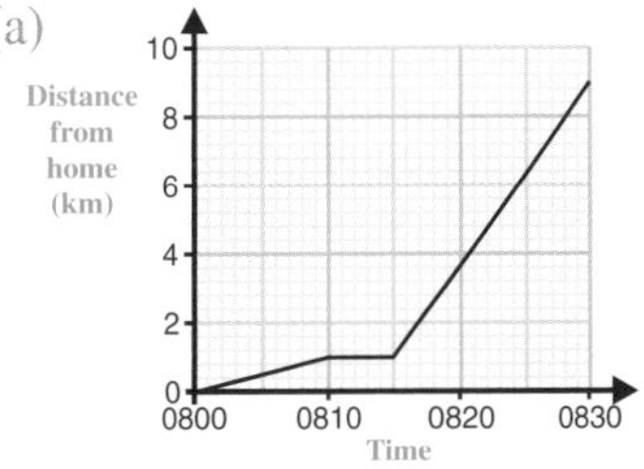

(b) 3.7 km

(c) (i) 36 minutes (ii) 0745

9. (a)

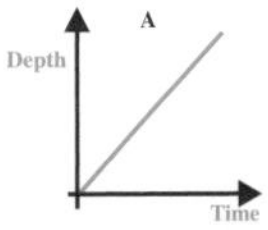

(b)

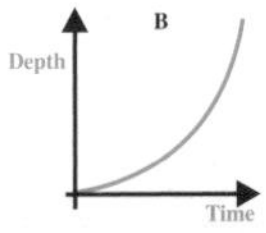

SECTION 19

Exercise 19

Page 45

1. (a) $x > 3$ (b) $x \geqslant -2$

(c) $x \leqslant 6$ (d) $x > 2$

2. (a)

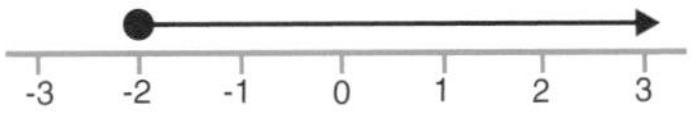

(b)

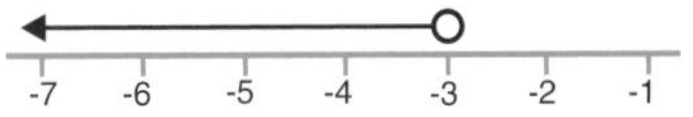

(c)

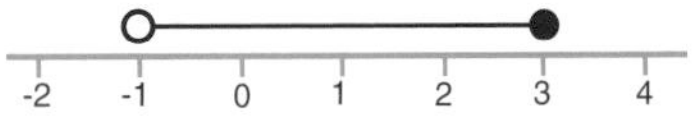

(d)

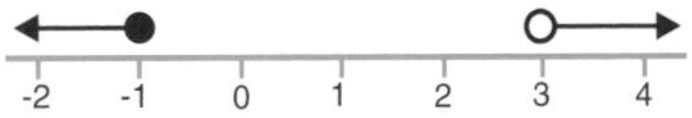

3. (a) $x \geqslant 3$

(b)

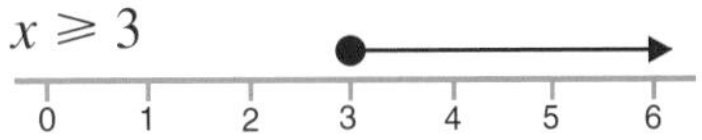

4. (a) $x \leqslant 2$ (b) $x > 3\frac{1}{2}$ (c) $x < -\frac{4}{3}$

5. (a) -1, 0, 1, 2 (b) 1, 2 (c) -1, 0, 1

6. -2, -1, 0, 1, 2, 3, 4

7. (a) $x \leqslant 2$ (b) $x > -1$ (c) 0, 1, 2

SECTION 20

Exercise 20

Page 46

1. (a) Missing entries are: 7, -1, 2

(c) $x = \pm 2.2$ (d) $x = \pm 1.4$

2. (a) Missing entries are: 2, −2, 2

(b) 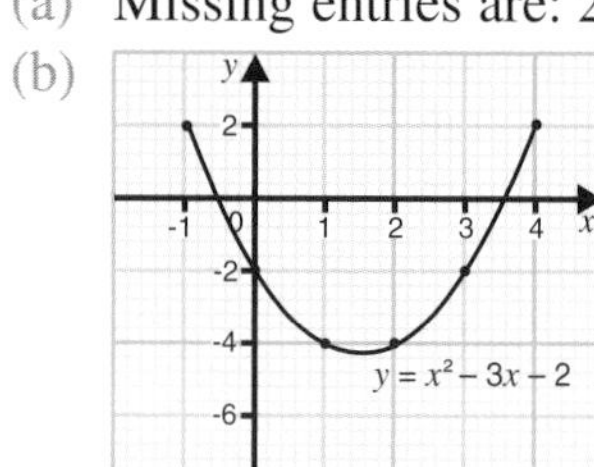

(c) $y = -4.25$

(d) $x = -0.6$ or 3.6

3. (b) $x = -0.7$ or 2.7

Algebra

Non-calculator Paper — Page 47

1. $C(1, 2)$, $D(-2, 3)$
2. (a) (i) 26 (ii) Add 4 to the last term.
   (b) 25, halve the last term.
3. (a) 48 (b) 7
4. 26 points
5. (a) (b) Missing entry is: 14
   (c) 17 sides
   (d) 35 sides
6. 21
7. (a) (b) $(-1, 1)$
8. (a) $5t$ pence (b) $(t + 5)$ pence
9. (a) (i) ▲ = 14 (ii) ■ = 5
   (b) (i) 29 (ii) Take 2 from the last term.
10. (a) **could be even or odd** (b) **always odd**
11. 10
12. $(3x + 5y)$ pence
13. (a) $10x - y$ (b) $35 - 10x$
14. $3a$ and $2a + a$ $2(a - 1)$ and $2a - 2$
15. (a) 8, −2 (b) $C = 6$
16. (a) $5x$ (b) $3a - 4b$ (c) $3m^2$
17. (a) $x = -3$ (b) $x = 2.5$
   (c) $x = 6$ (d) $x = 3$
18. (a) (i) y values are: −1, 2, 5, 8, 11
   (b) (i) Missing values are: −1, 3, 15
   (ii)
19. (a) $(x - 3)$ years (b) $4x$ years
   (c) $x + (x - 3) + 4x = 45$, $x = 8$.
   Louisa 5 years, Hannah 8 years, Mother 32 years.
20. $S = -112$
21. (a) 30 minutes (b) 18 km (c) 36 km/h
22. (a) (i) $3(a - 2)$ (ii) $k(k - 2)$
   (b) £$25x$
23. (a) $x = 3.5$ (b) $x = 6$ (c) $x = 7$
24. $t = 5$
25. (a) $x^2 + 2x$ (b) $16x + 1$
   (c) (i) $x \geqslant 4.5$
   (ii)
26. (a) $x = 1.5$ (b) $x = 3$
27. (a) 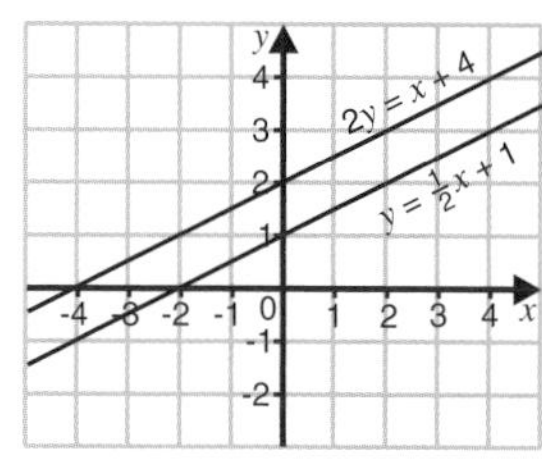

   (b) Lines are parallel, same gradient.
28. (a) $300 - x$
   (b) $15x + 10(300 - x) = 3950$, $x = 190$.
   So, 190 people paid £15.
29. (a) $3n + 1$ (b) 205
30. (a) $x = 2.5$
   (b) (i) 1, 2, 3, 4 (ii) $x \geqslant 0.6$
31. **A**: **Q**, **B**: **S**, **C**: **R**, **D**: **P**
32. (a) −3, 1 (b) $4n - 7$
33. (a) Missing entries are: 8, 3, −1
   (c) (i) $x = 1$ or 3
   (ii) $x = -0.4$ or 4.4
34.
35. (a) $x = 4.5$
   (b) $16x - 6$
36. (a) $p^2 - 4$ (b) q^6

Algebra

Calculator Paper — Page 50

1. (a) (i) 14 (ii) 15 (iii) 13
   (b) Take 4 away from the last term.
2. (a) $8a$ (b) $6x + 3y$
3. £34
4. (a) 14 (b) 8
5. (a) $3x$ pence (b) $(x + 30)$ pence
6. (a) (i) 30
   (ii) 50th term is an odd number. All even terms are odd numbers.
   (b) 12
7. 22. (Pattern number × 2) + 2
8. (a) (i) 25 dollars (ii) 18 euros
   (b) From graph, 20 dollars = 24 euros.
   So, 200 dollars = 10 × 24 = 240 euros.
9. (a) 3, −1
   (b) (i) 302 (ii) (3 × Position number) + 2

10. (a) Missing entries are: -3, 1
(b) (c) $(0, -2)$, $(2, 0)$

11. (a) £18.50 (b) £16

12. (a) $g = 8$ (b) $a = 5$
(c) $x = 6$ (d) $x = 3$

13. (a) $(5x + 8y)$ miles (b) 3

14. (a) $m = 3$ (b) $P = 45$

15. (a) 15 minutes (b) 0.75 km
(c) They increased speed. (d) 14 km/h

16. (a) 3 (b) Tom calculated:
$(2 \times 5)^2 + 1 = 100 + 1 = 101$
Correct calculation:
$2 \times 5^2 + 1 = 2 \times 25 + 1 = 51$

17. (a) (i) $a = 3.5$ (ii) $t = -1$
(b) $x + x - 3 + x + 7 = 25$
$3x + 4 = 25$, $x = 7$.

18. (a) £$(227 + 9n)$
(b) $227 + 9n = 299$, $n = 8$
Pali worked 8 hours overtime.

19. (b) $P(-5, -9)$

20. (a) (i) $5(2x + 3)$ (ii) $x(x - 3)$
(b) $x = 1.7$

21. (a) -2, -9 (b) $3n + 2$

22. $x = 2.66$.

23. (a) Missing entries are: -1, -4, -1
(c) $x = \pm 2.2$

24. (a) $14x + 7$ (b) $L = \frac{P - 2W}{2}$

25. (a) $(x + 45)$ pence
(b) $3(x + 45) + x = 455$, $x = 80$.
Glass of milk costs 80 pence.

26. (a) $x > 3$
(b) -3, -2

27. (a) a^7
(b) $t = \frac{P - 5}{3}$

28. $x = 4.2$

29. (a) $x = -\frac{1}{5}$
(b) $7x - 3$
(c) (i) m^6 (ii) n^5

SECTION 21

Exercise 21 Page 54

1. (a) $\angle ABC = 25°$ (b) Acute angle

2. (a) CD and EF (b) AB and CD
(c) (i) $y = 135°$ (ii) **obtuse angle**

3. (b) $m = 96°$ (supplementary angles)

4. (a) $a = 143°$ (supplementary angles)
(b) $b = 135°$ (angles at a point)
(c) $c = 48°$ (vertically opposite angles)
$d = 132°$ (supplementary angles)
$e = 44°$ ($3e = 132°$, vert. opp. angles)

5. (a) $x = 53°$ (alternate angles)
(b) $y = 127°$ (allied angles or supplementary angles)

6. $x = 36°$

7. $a = 68°$ (supplementary angles)
$b = 112°$ (corresponding angles)
$c = 106°$ (allied angles)

8. (a) $a = 105°$ (b) $b = 117°$, $c = 117°$
(c) $d = 42°$, $e = 76°$, $f = 62°$

SECTION 22

Exercise 22 Page 55

1. (a) $a = 27°$ (b) $b = 97°$ (c) $c = 125°$

2. $a = 125°$ (supplementary angles)
$b = 82°$ (sum of angles in $\Delta = 180°$)

3. $x = 130°$ ΔPQR is isosceles, $\angle PQR = \angle PRQ$

4. $a = 50°$ (vertically opposite angles)
$b = 80°$, $180° - 2 \times 50°$

5. (a) $x = 64°$ (b) $y = 122°$

6. (b) (i) $a = 22$ cm (ii) $x = 54°$ (c) 117 cm^2

8. (b) 15.2 cm^2

9. 14 cm^2

10. $YX = 4$ cm

SECTION 23

Exercise 23 Page 58

1. (a) (b) 2

2. (a) A, E (b) N
(c) O

3. (a) (i) 3 (ii) 3
(b) (i) 0 (ii) 1
(c) (i) 0 (ii) 4
(d) (i) 1 (ii) 1

4. (a) (i) **Tray**
(ii) **Jug**
(b) E.g.

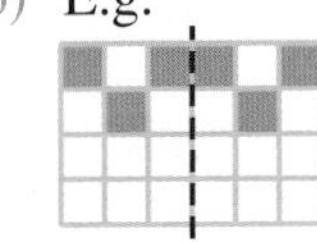

5. (a) 4 (b) 1

6. **A** and **F**

7. **B** and **D** (SAS)

SECTION 24

Exercise 24 Page 60

1. (a) 14 cm^2
(b) (i)

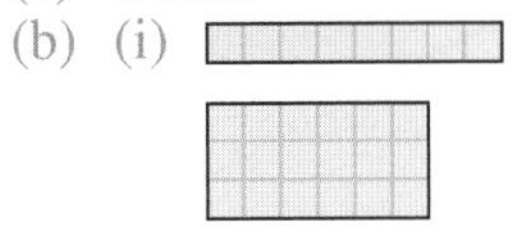

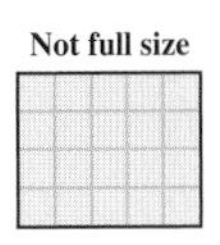

Not full size

(ii) 8 cm^2, 18 cm^2, 20 cm^2

2. (a) **square**, **rhombus**
(b) E.g. Opposite sides are equal and parallel.

3. (a) $a = 70°$ (b) $b = 132°$
(c) $c = 110°$, $d = 120°$

4. (a) **kite** (b) $\angle ABC = 114°$

5. £280

6. (a) $20 \times 10 = 200$ m^2
(b) Bigger, both dimensions rounded up.

7. (a) (i) $x = 70°$ (supplementary angles)
(i) $y = 35°$, $z = 110°$
(b) 60 cm^2

SECTION 25

Exercise 25 — Page 61

1. Shape A has rotational symmetry of order 6 and 6 lines of symmetry.
   Shape B has rotational symmetry of order 2 and 2 lines of symmetry.
2. $x = 72°$
   $\angle AEB = \frac{(180° - 108°)}{2} = 36°$, isosceles Δ.
   $x = 108° - 36° = 72°$
3. (a) $a = 53°$ (b) $b = 115°$ (c) $c = 140°$
4. (a) $a = 120°$ (b) $b = 60°$, $c = 120°$
   (c) $d = 72°$, $e = 108°$
5. (a) The shapes cover a surface without overlapping and leaving no gaps.
   (b) 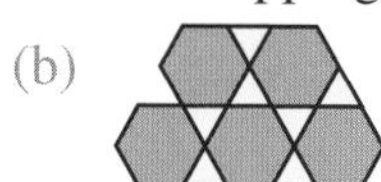
6. (a) $\angle ABC = 144°$ (b) $\angle XCY = 108°$
7. 15 sides
8. 140°
9. $\angle PQX = 151°$
10. Number of sides $= \frac{360°}{30°} = 12$
    Sum of interior angles $= (12 - 2) \times 180°$
    $= 1800°$

SECTION 26

Exercise 26 — Page 63

1. South-East
2. (a) Montego Bay (b) Port Antonio
   (c) East
3. 26 cm
4. (a) 3.5 km
   (b) 029°
   (c) 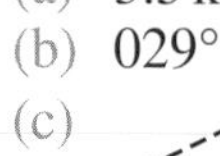
   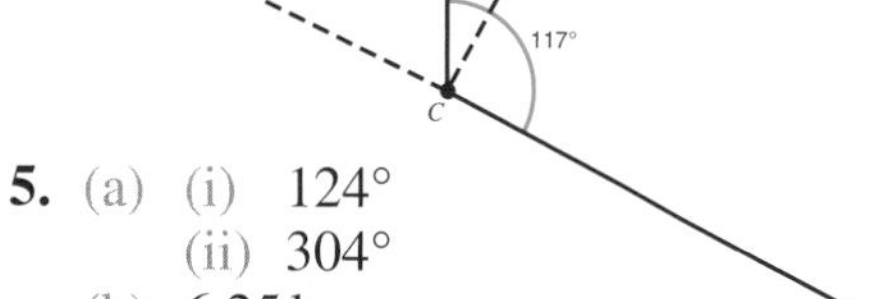

5. (a) (i) 124°
   (ii) 304°
   (b) 6.25 km
6. (b) 135°
7. (b) (i) 250° (ii) 1530 m

SECTION 27

Exercise 27 — Page 66

1. (a) (i) 1 cm (ii) 3.14 cm^2 (b) 6.28 cm
2. (a) $C = \pi d = 3 \times 10 = 30$ m
   (b) $A = \pi r^2 = 3 \times 5 \times 5 = 75\ m^2$
3. (a) 15.7 cm (b) 19.6 cm^2
4. $225\pi\ cm^2$
5. 754 cm^2
6. 19.9 times
7. (a) 346 cm (b) 7977 cm^2
8. 27.5 m^2
9. 61.7 cm
10. 796 cm^2
11. Yes. Semi-circle $= \frac{1}{2}(\pi \times 10^2) = 50\pi\ cm^2$
    Circle $= \pi \times 5^2 = 25\pi\ cm^2$

SECTION 28

Exercise 28 — Page 68

1. (a) 5 faces, 8 edges, 5 vertices (b) **R**
2. (a) 18 cm^2 (b) 20 cm^2 (c) 22 cm
3. 13 cm^2
4. (a) 30 cubes (b) (i) 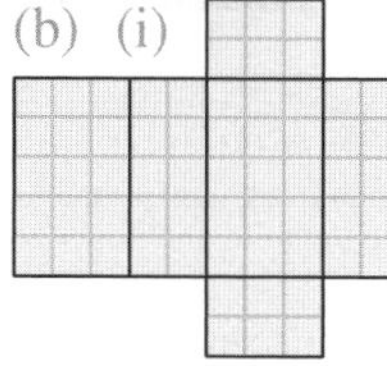 (ii) 62 cm^2
5. (a) (i) 6 (ii) **B**: 2, **C**: 3
   (b) (i) 3 cm^3 (ii) 18 cm^3
6. (a) (b) 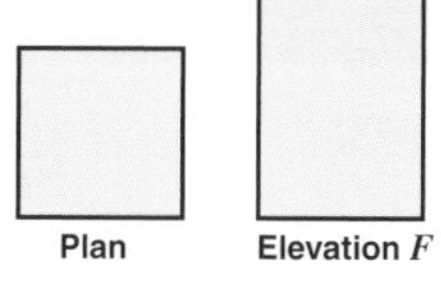
7. **C**. **A** $= 24\ cm^3$, **B** $= 24\ cm^3$, **C** $= 27 cm^3$
8. (a) 1.08 m^3 (b) 0.78 m^3
9. (a) 9.4 cm (b) E.g. 2 cm by 3 cm by 6 cm
10. 86 cm^2
11. 297 m^2
12. (a) 414 cm^2 (b) 405 cm^3
13. 110 000 cm^3
14. 26.7 cm^3
15. 424 cm^3
16. (a) 1.9 m^3 (b) 1900 litres
17. (a) 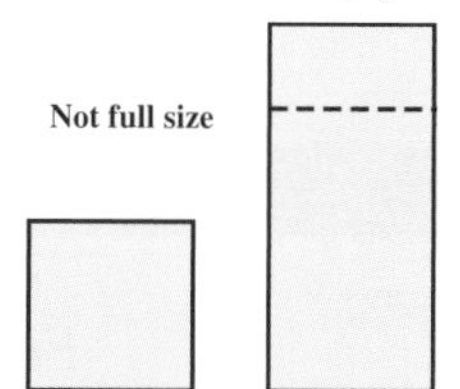

    (b) 55 600 cm^3
18. (a) 32 673 cm^2 (b) 20.1 cm

SECTION 29

Exercise 29 — Page 72

1. 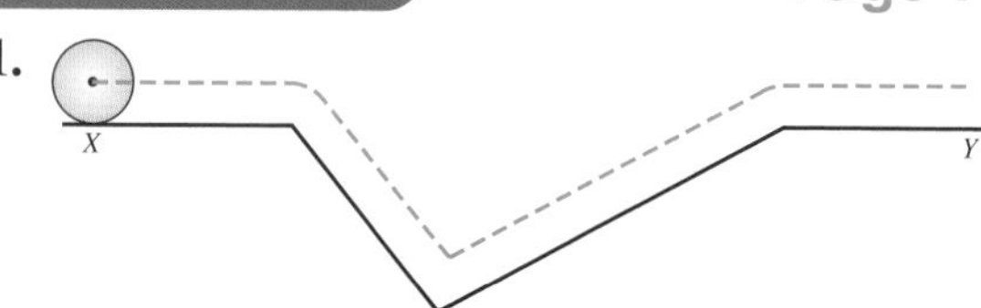

2. (a) (b)

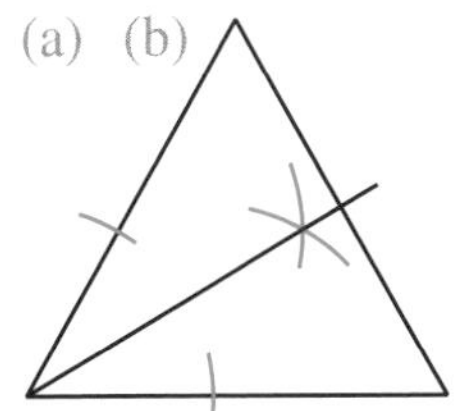

4. (a) (b)

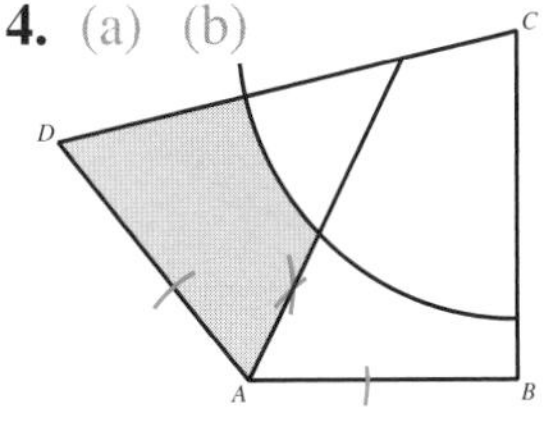

3.

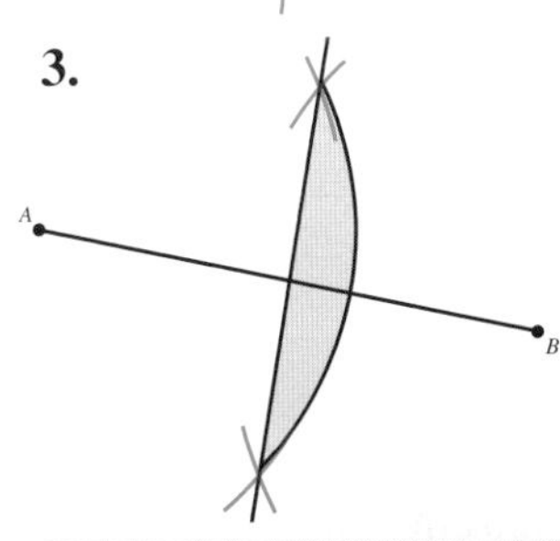

5. (a) (b)

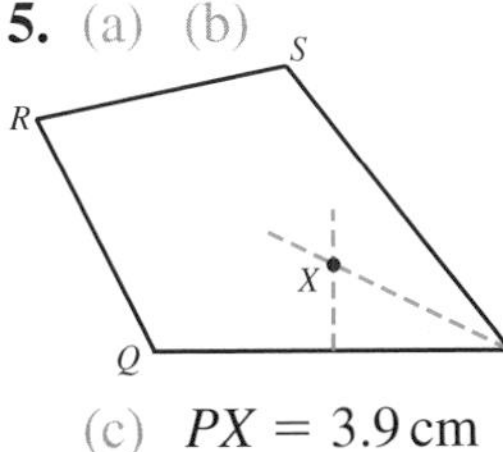

(c) $PX = 3.9$ cm

SECTION 30

Exercise 30 — Page 74

1. (a)

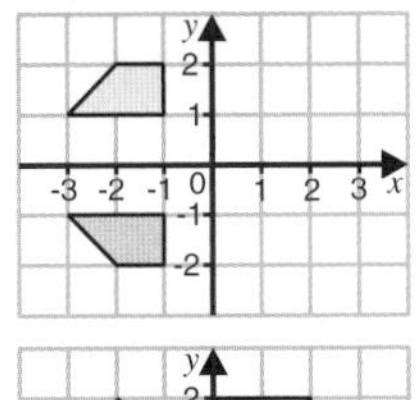

(b)

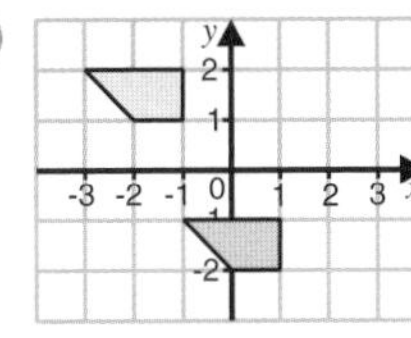

(c)

2. (a) Rotation, 90° anticlockwise, about (0, 0).
(b) Translation 4 units to the right and 3 units down $\left(\text{or} \begin{pmatrix} 4 \\ -3 \end{pmatrix}\right)$.
(c) Reflection in $y = 3$.

3. (a)

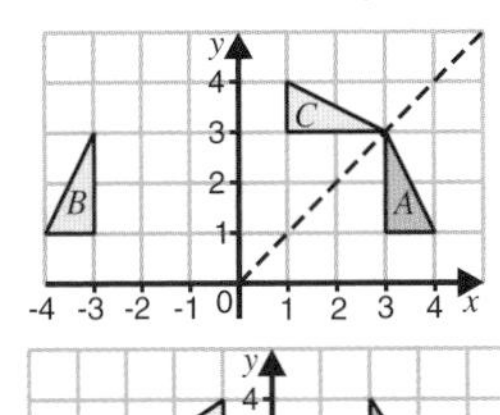

(b)

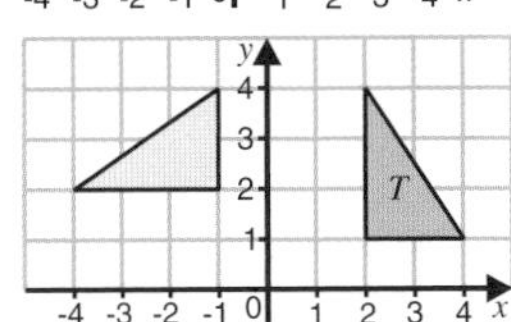

(c)

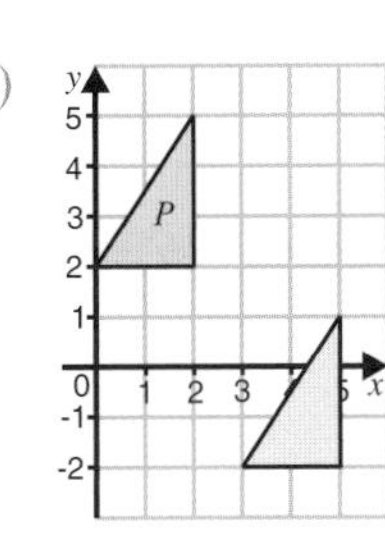

4. (a) $x = -1$
(b) One unit to the left and 2 units up.
(c)

5. (a) (i) Reflection in $x = 2$.
(ii) Translation $\begin{pmatrix} -6 \\ -5 \end{pmatrix}$.
(b) Coordinates of triangle R: (0, −3), (2, −3), (0, −4).

6. (a) (i) Rotation, through 90° clockwise, about (0, 0).
(ii) Reflection in $y = x$.
(b) Coordinates of triangle D: (−3, 5), (−3, 3), (−2, 5).

7. (a) Reflection in $x = 3$.
(b) Rotation, through 180°, about (2, 1).
(c) Translation $\begin{pmatrix} 2 \\ -3 \end{pmatrix}$.

8. (a) Translation $\begin{pmatrix} 3 \\ 2 \end{pmatrix}$.
(b)

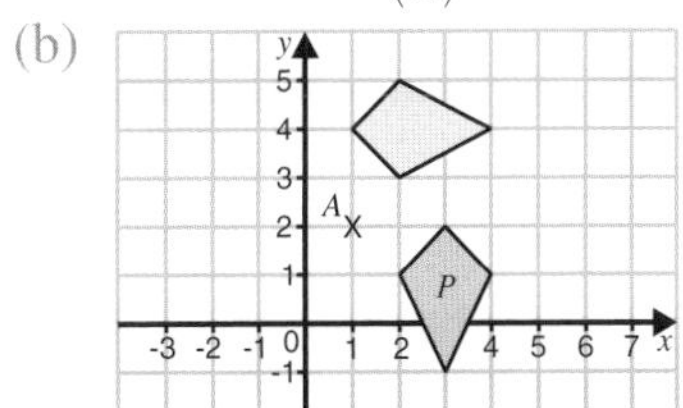

9. (a) (i) Reflection in the y axis ($x = 0$).
(ii) Rotation, through 90° anticlockwise, about (0, 0).
(b) Rotation, through 180°, about (4, 1).

SECTION 31

Exercise 31 — Page 77

1.

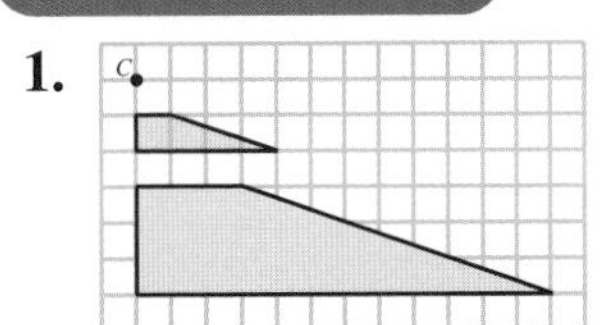

2.

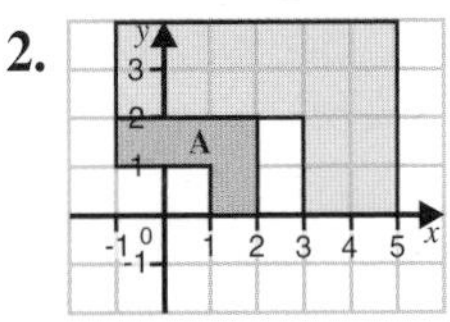

3. Enlargement, scale factor 3, centre (0, 0).

4. (a) Centre (0, 2), scale factor 2.
(b) Coordinates of enlarged shape: (3, 2), (3, 4), (1, 4), (−1, 0).

5.

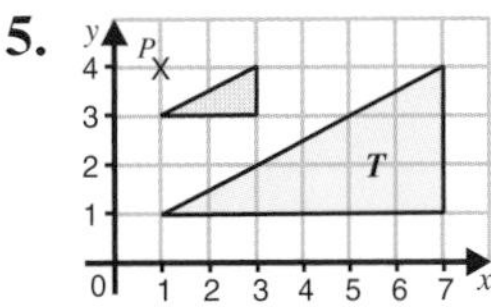

6. (a) Ratio of widths = 1 : 2 **but** ratio of lengths = 5 : 7. (b) 7.5 cm

7. (a) $x = 7.8$ cm (b) $y = 4.3$ cm

SECTION 32

Exercise 32 — Page 78

1. $BC = 13$ cm
2. $PQ = 8$ m
3. 28.7 cm
4. (a) (3, 5) (b) 7.21 units
5. 48.8 m
6. $QR = 7.48$ cm; Area $\Delta PQR = 18.7\ \text{cm}^2$
7. $AB = 5.6$ cm
8. $AC = 4.47$ m
9. $AC = 10$ cm
10. $AE = 5.7$ cm

SECTION 33

Exercise 33 — Page 81

1. (a) **15 grams** (b) **3.28 kg** (c) **44 km/h**
2. (a) Square metres (b) Grams (c) Kilometres (d) Litres
3. 2 mm, 20 mm, 20 cm, 2 m

4. $\frac{1}{4}$

5. (a) (i) 27 m
(ii) 4 m
(b) 1800 grams

6. 14 000 cm^3

7. (a) 1.8 m^2
(b) 200 000 cm^3

8. £813

9. 165 mm

10. 2.45 kg

11. 24 km/h

12. $\frac{\pi a^2}{4} + \frac{\pi ac}{2}$
has dimension 2.

Shape, Space, Measures

Non-calculator Paper

1. (a) 7.7 cm
(b) - (f)

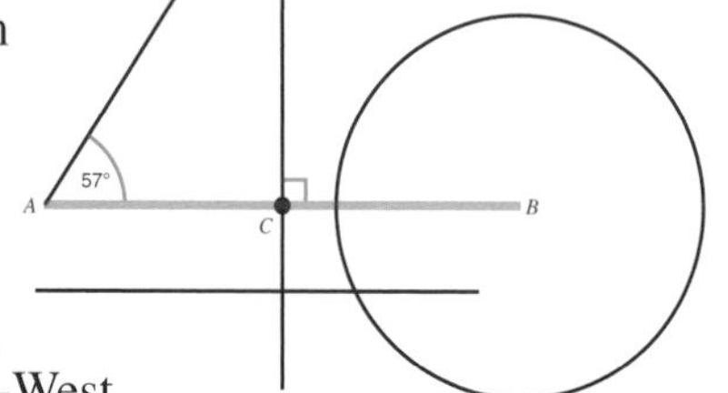

2. (a) South
(b) North-West

3. (a) 12 (b) 5 (c) cone

4. 22 cm^2

5. (a) 60 boxes (b) 1.2 m by 1.6 m by 2 m

6. (a) 2 (b) Congruent
(c) (i) 5 cm^2 (ii) 20 cm^2
(d) $\begin{pmatrix} 0 \\ -3 \end{pmatrix}$ or 3 units down.

7. (a) $x = 34°$, $y = 110°$ (b) y

8. (a) $a = 58°$ (vertically opposite angles)
(b) $b = 155°$ (supplementary angles)
(c) $c = 150°$ (angles at a point)

9. (a) (i) 5 (ii) 3
(iii) 2 (c)
(b) E.g.

10. (a) (b)

11. (a) $a = 67°$ Angles in a triangle add to 180°.
(b) $b = 54°$ Isosceles Δ.
$b = 180° - (2 \times 63°)$
(c) $c = 126°$ Angles in a quadrilateral add to 360°.
$c = 180° - (360° - 306°)$

12. (a) 2500 cm (b) 25 m

13. $140° + 40° = 180°$, so, allied angles.
So, AB is parallel to CD.

14. (b) 308°

15. (a) (i) 2400 cm^3 (ii) 2.4 litres (b) 5 cm

16.

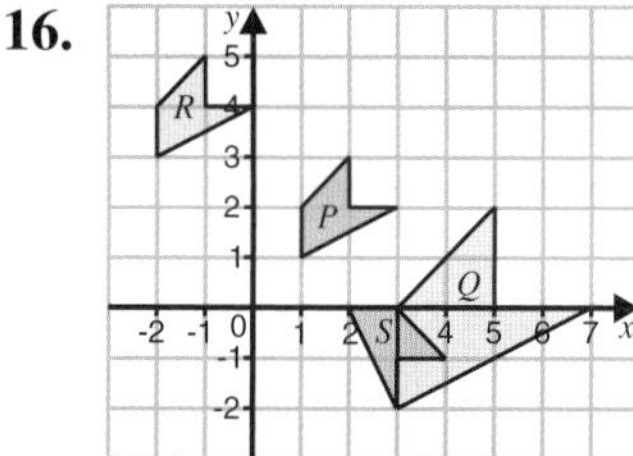

17. (a) $x = 52°$ (b) $y = 85°$ (c) $z = 21°$

18.

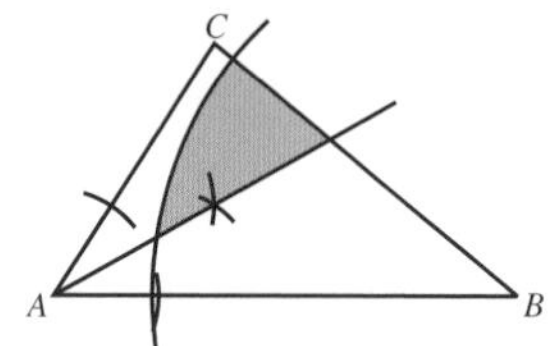

19. 3.6 cm

20. 14π cm

21. 2.25 m

22. (a) abc has dimension 3 (volume).
(b) πa **and** $\sqrt{a^2 - c^2}$ **and** $2(a + b + c)$

23. (a) (i) $x = \sqrt{3^2 + 4^2} = \sqrt{9 + 16} = \sqrt{25} = 5$
(ii) 174 m^2
(b) $y = 8$ m

Shape, Space, Measures

Calculator Paper

Page 85

1. (a) 8000 g and 8 kg
(b) 0.2 km
(c) 1.4 kg

2. **Square**: E
Trapezium: C
Parallelogram: B
Rhombus: F

3.

4. (a) $A(2, 3)$
(b) $C(-2, -1)$
(c) $D(-2, 3)$

5. (a) 2.8 cm
(b) (c)
(d) 314 cm

6. Height 175 cm
Weight 70 kg

7. (a) 40 cm^3 (b) 20 cm^2

8. $30° + 40° + 70° \neq 180°$
Sum of angles in a triangle should be 180°.

9. **200 ml, $\frac{1}{2}$ litre, 0.7 litre, 2 litres**

10. (a) Reflection in the line $x = 5$.
(b) Rotation, 90° anticlockwise, about (0, 0).
(c) Translation $\begin{pmatrix} 4 \\ 1 \end{pmatrix}$

11. 27.75 cm^2

12. (a) E.g.

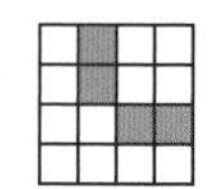

(b) E.g.

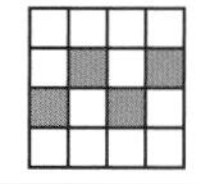

13. (a) E.g.

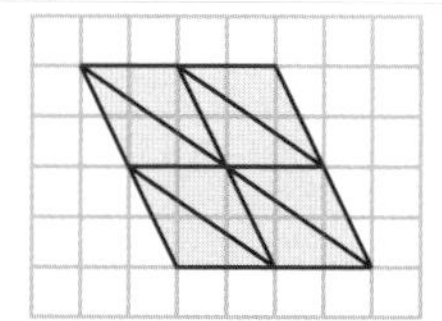

(b) No. Only equilateral triangles, squares and hexagons tessellate.
Interior angle must divide into 360° a whole number of times.

14. (a) (i)

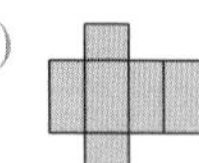

(ii) 46 cm^2
(b) 2.5 cm

15. (a) (i) $x = 41°$
(ii) $y = 139°$ (supplementary angles)
(b) (i) $z = 142°$ (allied angles)
(ii) 147.5 cm^2

16. 50.3 cm^2

17. E.g.

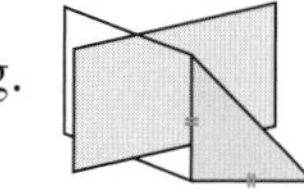

18. (a) Ext. $\angle = \frac{360°}{6} = 60°$.
$a = \text{int.} \angle = 180° - 60° = 120°$
(b) $b = 15°$

19. 36.25 km

20. (a)

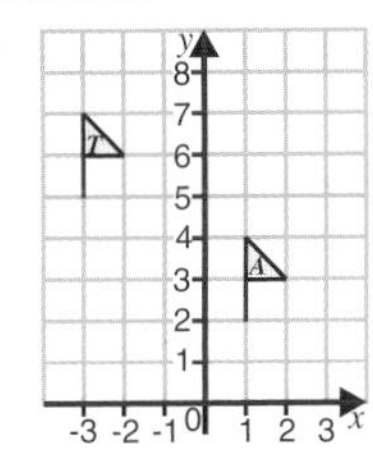

(b) (i) Enlargement, scale factor 2, centre (0, 0).
(ii) Rotation, 90° clockwise, about (3, 2).

21. (a) 41 m² (b) (i) AB = 5 m (ii) 30 m

22. (b) (i) 215° (ii) 17 km

23. Footpaths HX, XS, shorter by 171 m.
Footpaths 850 m, Waverly Crescent 1021 m.

24. 280 000 cm³

SECTION 34

Exercise 34 — Page 88

1. (a) (i) E.g.

Gender	Mode of transport
M	Bus
F	Cycle

(b) Data will be biased, as people at bus station are likely to travel by bus.

2. (a) Laila (b) Ria
(c) Corrin. Pupils are in the same class and June is a later month in the school year than the other months given.

3. E.g.

Resort	Tally
Cervinia	卌 II
Livigno	卌 II
Tonale	卌 IIII

4. (a)

Number of absences	Tally	Frequency
0 - 4	卌 III	8
5 - 9	卌 II	7
10 - 14	IIII	4
15 - 19	卌 I	6
20 - 24	卌	5
25 - 29	II	2

(b) (i) 32 (ii) 25

5. (a)

	Walk	Bus	Cycle	Totals
Boys	1	4	8	13
Girls	9	8	0	17
Totals	10	12	8	30

(b) 1 boy walks to school.

6. (a) (i) Too personal.
(ii) In which age group are you?
Under 16 ☐ 16 to 19 ☐ Over 19 ☐
(b) (i) Only students already using the library are sampled.
(ii) Give to students as they enter (or leave) the college.

7. E.g. Leading question.
Question has more than one part.

8. E.g. Two thirds of men were over 45.
All women are aged 16 to 45.
Twice as many women as men.

9. No. Men: $\frac{180}{200} = 90\%$ Women: $\frac{240}{300} = 80\%$
Higher proportion of men can drive.

10. (a) 2 (b) Yes. 21 people have cats **and** 17 people have dogs.
(c) 25 dogs

SECTION 35

Exercise 35 — Page 91

1. (a)

Vowel	Tally
A	卌 IIII
E	卌 III
I	卌 III
O	卌 I
U	IIII

(b)

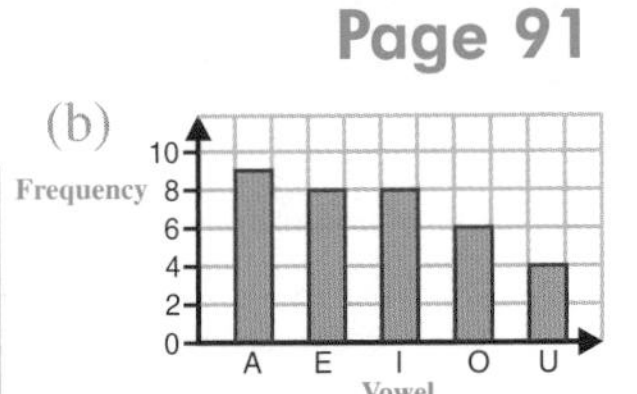

(c) *A*

2. (a) 20 (b) 35 (c)

3. (a) Wednesday (b) 3 hours (c) 39 hours

4. (a) 8 (b) 16%
(c)

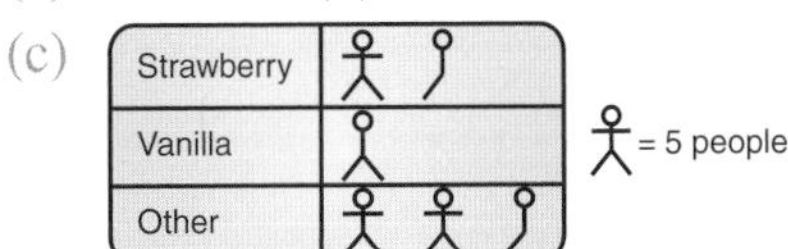

5. (a) Range: 2, mode: 2.
(b) Women's team had a larger range (5) and higher mode (3).

SECTION 36

Exercise 36 — Page 94

1. (a) 7 cm (b) 10 cm (c) 11 cm (d) 11.3 cm

2. (a) 13 (b) 8 eggs

3. (a) £9 (b) £10.50 (c) £11.50
(d) Median. Mode is the lowest price and mean is affected by the one higher-priced meal.

4. (a) 5 (b) 6

5. (a) (i) 4.9 hours (ii) 6 hours
(b) Much bigger variation in the number of hours of sunshine each day and lower average.

6. 99.5 kg
7. (a) … the mean and median distances are greater.
   (b) … the range for college students is 12 km, but the median for school students is 0.75 km and the range is only 5.5 km.
8. (a) 5 (b) 9
   (c) Reg: mean; Reg 9, Helen 8.6
   Helen: mode; Reg 9, Helen 10
   Friend: median; Reg 9, Helen 9
9. (a) (i) 1 (ii) 3 (iii) 3.35
   (b) Mode **and** median
10. (a) £23.20
   (b) The calculation is based on the average for each class interval, which is at the centre of the class.
11. (a) 6.069 hours
   (b) No. $3 \times 2.08 = 6.24$ hours.
   So, *Megagum* flavour lasts less than 3 times average for other brands.

SECTION 37

Exercise 37 Page 96

1. (a) 12 (b) 22°C (c) 14°C
   (d) 5°C or 24°C. Temperature can be either 2°C above previous maximum, or 2°C below previous minimum.
2. (a) $\frac{1}{4}$ (b) 30° (c) 13 students
3. (a)

Sport	Soccer	Rugby	Cricket	Basketball	Other
Angle	148°	50°	36°	74°	52°

   (b) 21 girls
4. (a) 1 | 0 means 10 text messages

0	2 3 5 5 7 9
1	0 1 2 3 5 7
2	0 1

   (b) 19
5. (a) $8\frac{1}{3}\%$ (b) ***Sometimes*** (c) $\frac{1}{4}$
   (d)

Response	*Never*	*Sometimes*	*Quite often*	*Very often*	*Always*
Angle	68.5°	162°	79°	32.5°	18°

   (e) E.g. A higher proportion of boys said they always considered their health than girls.
6. (a) 2 | 5 means 2.5 cm

Boys		Girls
	2	5
5 5	3	0 5 5 5
5 5 5 5 0 0	4	0 5 5
0 0	5	0 5

   (b) Girls have more variation in their estimates. Range: girls 3 cm, boys 1.5 cm

SECTION 38

Exercise 38 Page 99

1. (a)

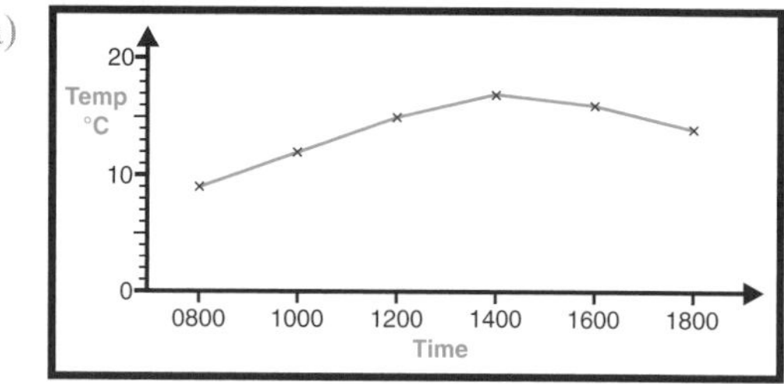

   (b) 8°C
   (c) (i) 16°C
   (ii) Actual temperatures are only known at times when readings are taken.
2.

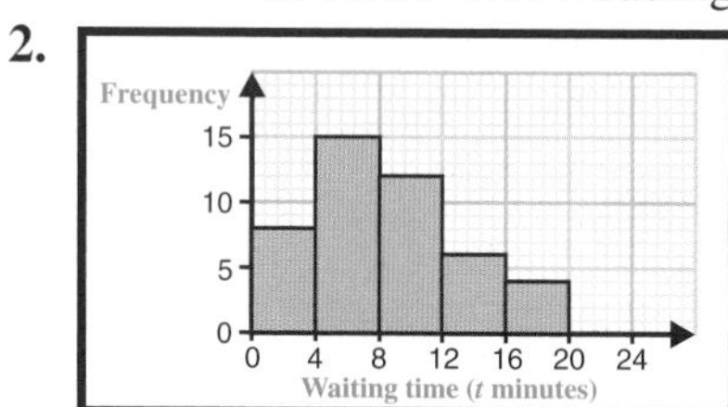

3. (a)

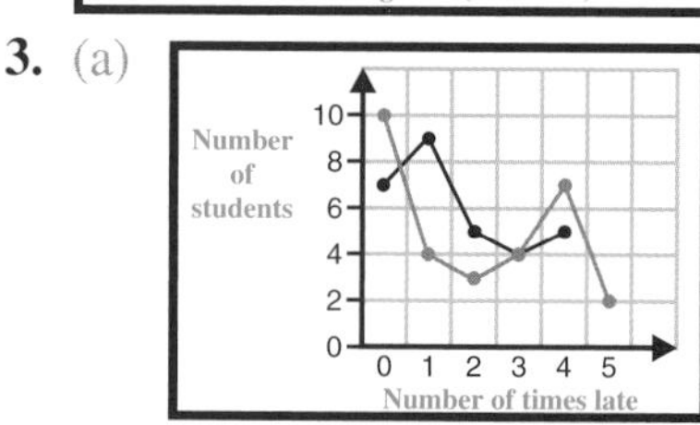

   (b) (i) 10A: 1, 10D: 0 (ii) 10A: 4, 10D: 5
   (c) 2
   (d) E.g. Mean similar for both forms.
4. Vertical scale does not begin at zero.
   Bars are different widths.
5. (b) 5%
6. (a) $80 \leqslant age < 90$ (b) 40
   (c) (ii) Women:

Age (a years)	Frequency
$60 \leqslant a < 70$	1
$70 \leqslant a < 80$	5
$80 \leqslant a < 90$	13
$90 \leqslant a < 100$	6

   (iii) More men under 80 than women.
   Only women aged over 90.
   Women have greater range of ages.
7. (a) 7 (b) 3 (c) 67

SECTION 39

Exercise 39 Page 101

1. (b) ***Strong, negative*** (d) 10 hours
2. (a) (i) ***C*** (ii) ***A*** (iii) ***B*** (b) ***C***
3. (a) 36 (b) 60 (c) No
   (d) Yes (e) Positive

4. (b) Negative correlation.
Members who spend longer on exercise each day tend to have a lower resting pulse rate.
(d) Estimate would be beyond know values.

SECTION 40

Exercise 40 Page 104

1. (a) **certain** (b) **evens** (c) **unlikely**

2. Number line from 0 to 1: C at 0, B at $\frac{1}{2}$, D at $\frac{3}{4}$ approx., A at 1.

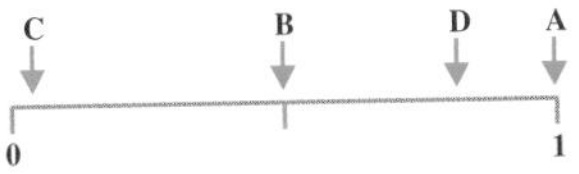

3. (a) $\frac{1}{50}$
(b) Only 9 cards have 1-digit, but 41 cards have 2-digits, so, higher probability of a 2-digit number.

4. (a) **Grey spinner** (across), **White spinner** (down)

	1	2	3
1	2	3	4
2	3	4	5
3	4	5	6
4	5	6	7

(b) $\frac{3}{12} = \frac{1}{4}$

5. (a)

G	B	C
G	C	B
B	G	C
B	C	G
C	B	G
C	G	B

(b) (i) $\frac{7}{20}$ (ii) 0.6

6. (a) 0.1 (b) 0.7

7. $\frac{20}{50} = \frac{2}{5}$ (or 0.4)

8. (a) $\frac{5}{12}$ (b) $\frac{5}{11}$

9. (a) $\frac{9}{20} = 0.45$
(b) 2, 3, 3, 4, 5. Numbers 2, 3, 4, 5 have occurred and 3 has occurred twice as often as other numbers.
(c) 100. Relative frequency of 5 is $\frac{1}{5}$.
$\frac{1}{5} \times 500 = 100$

10. (a) HHH, HHT, HTH, THH
THT, HTT, TTH, TTT
(b) $\frac{3}{8}$

Handling Data

Non-calculator Paper Page 106

1. (a) 4 (b) 11 (c) 22
2. (a) Spain (b) 12 (c) 51
3. (a)

<table>
<tr><th>Score</th><th>Tally</th><th>Frequency</th></tr>
<tr><td>2</td><td>|</td><td>1</td></tr>
<tr><td>3</td><td>|||</td><td>3</td></tr>
<tr><td>4</td><td>𝍸 ||</td><td>7</td></tr>
<tr><td>5</td><td>𝍸 ||||</td><td>9</td></tr>
<tr><td>6</td><td>𝍸 |||</td><td>8</td></tr>
<tr><td>7</td><td>𝍸</td><td>5</td></tr>
<tr><td>8</td><td>|||</td><td>3</td></tr>
</table>

(b)

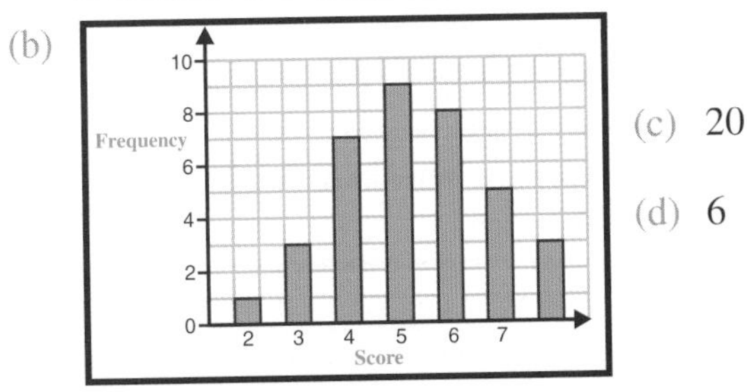

(c) 20
(d) 6

4. (a) **5** (b) **7**
5. (a) 3 (b) 30 (c) 60
6. (a) (i)

Pantomime	Angle
Aladdin	135°
Cinderella	105°
Jack and the Bean Stalk	75°
Peter Pan	45°

(ii) Aladdin
(b) (i) 72
(ii) $33\frac{1}{3}\%$

7. (a) **Second dice** (across), **First dice** (down)

×	1	2	3	4	5	6
1	1	2	3	4	5	6
2	2	4	6	8	10	12
3	3	6	9	12	15	18
4	4	8	12	16	20	24
5	5	10	15	20	25	30
6	6	12	18	24	30	36

(b) (i) $\frac{1}{36}$
(ii) $\frac{6}{36} = \frac{1}{6}$
(iii) 0

8. (a) 4 | 5 means 4.5 cm

Stem	Leaves
4	5 8 8
5	0 0 4 4 5 8
6	0 2 4 5 5 5 6 8
7	0 2 4

(b) 2.9 cm

9. E.g. Leading question.
Question has more than one part.

10. (a)

First	N	N	N	N	N	N
Second	A	A	R	R	C	C
Third	R	C	A	C	R	A
Fourth	C	R	C	A	A	R

(b) $\frac{1}{3}$

11. (c) Positive correlation.
Older people tend to have higher blood pressure.

12. (a) $\frac{17}{75}$ (b) Yes. Female: $\frac{12}{50} = 24\%$
Male: $\frac{5}{25} = 20\%$

13. (a) (i) 2 (ii) 5
(b) The next 25-week period had more variation in the number of accidents per week, but a lower weekly average.

14. (a) 0.7 (b) 0.44

Handling Data

Calculator Paper Page 109

1. (a) 170
(b)

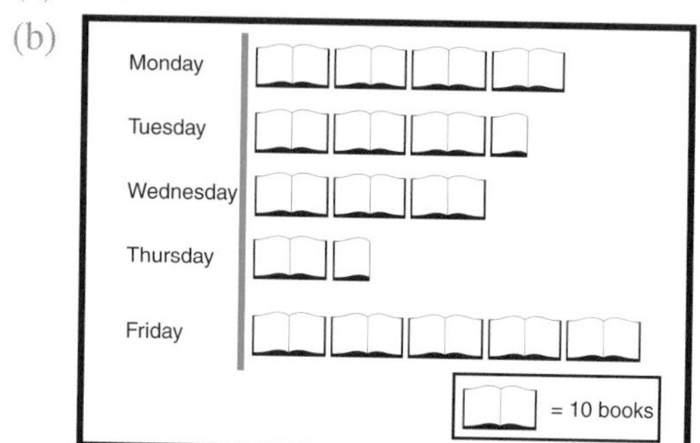

2. (a) 39°C (b) −14°C (c) 53 degrees
(d) 5°C (e) 7 hours

3. (a) **25.4 litres** (b) **23.9 litres**

4. (a) **X 1, X 3, Y 1, Y 3**
(b) Numbers 1 and 3 are not equally likely.

5. (a) 20 (b) 6 (c) 5
(d)

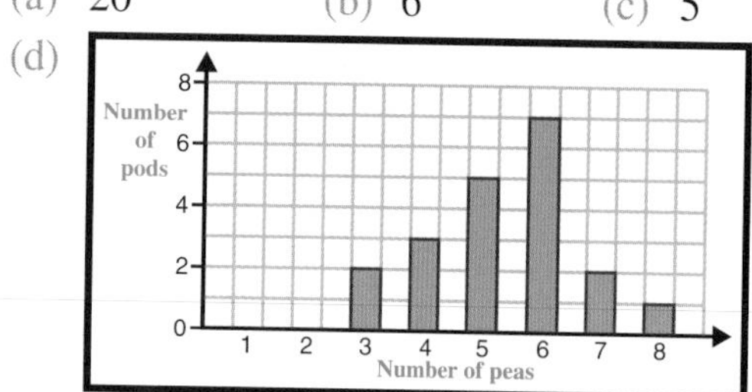

6. (a) 15 (b) 28 g (c) 30 g (d) 29.3 g

7. 0.25

8. (a) $\frac{3}{20}$ (b) $\frac{7}{20}$

9. (a)

Activity	Paid job	Doing homework	In bed	Sport	Other
Angle	130°	43°	90°	83°	14°

(b) (i) Similar proportion of students doing paid jobs.
(ii) Higher proportion of Year 11 stay in bed than Year 13.

10. (a) 0.8 (b) 50

11. 56.7 kg

12. (a) (i) 2 (ii) 1.9 (b) (i) 0.75 (ii) 12

13. (b) Circle point (2.5, 10)
(c) Negative correlation, fairly strong.
(d) (ii) 9.5 km

14. 72

15. (a) 178.5 seconds
(b) $150 < t \leqslant 180$
There are 40 tracks, and the 20th and 21st times occur within the class interval.
(c) $\frac{3}{40}$

Exam Practice

Non-calculator Paper Page 112

1. (a) 2609 (b) Sixty million

2. (a) 3 (b)

Monday	⊕ ⊕	⊕ = 4 cars
Tuesday	⊕ ⊕ ⊕	
Wednesday	⊕ ◖	
Thursday	⊕ ⊕ ◔	
Friday	⊕ ⊕ ◖	

3. (a) **7, 19, 30, 105, 2002**
(b) (i) 75 (ii) 133 (iii) 286

4. (a) Centimetre (b) Gram
(c) Kilometre (d) Millimetre

5. (a) Add 6 to the last number, 33.
(b) Double the last number, 32.

6. (a) (i) 2 (ii) 1 (b) 16 cm (c) 8 cm^2

7. (a) $5d$ pence (b) $(d + 25)$ pence

8. (a) (i) 229 (ii) 1300 (iii) 0.132
(b) (i) Any two of: 1, 3, 7, 21
(ii) 21, 42, 63, …
(iii) 25

9. (a) £192 (b) 12 coaches

10. (a) 3 m (b) 35 cm

11. $P = 26$

12. (a) $\frac{3}{10}$ (b) 70%

13. (a) £37 (b) 15 days

14. (b) (3, −2)

15. (a) (i) (ii) (iii) *None*
(b) 5 (c)

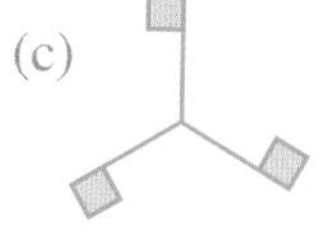

16. (a) 59 pence (b) 49 pence (c) 50 pence

17. −2°C

18. $a = 99°$

19. (a) $9p + 2q$ (b) (i) $w = 3$ (ii) $z = 3.5$

20. (a)

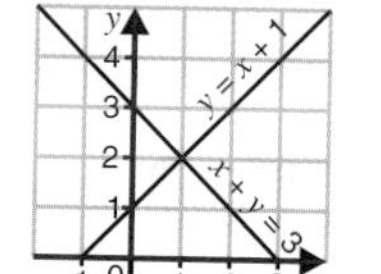

(b) (1, 2)

21. $\angle BAC = 38°$

22. (a) E&C, RP; E&C, G; P&M, RP; P&M, G
C&R, RP; C&R, G; F&C, RP; F&C, G

(b)

Dish	Egg	Pie	Curry	Fish
Angle	40°	76°	80°	164°

23. (a) 34 (b) 25

24. (a) 15 (b) $\frac{1}{8}$

25. (a)

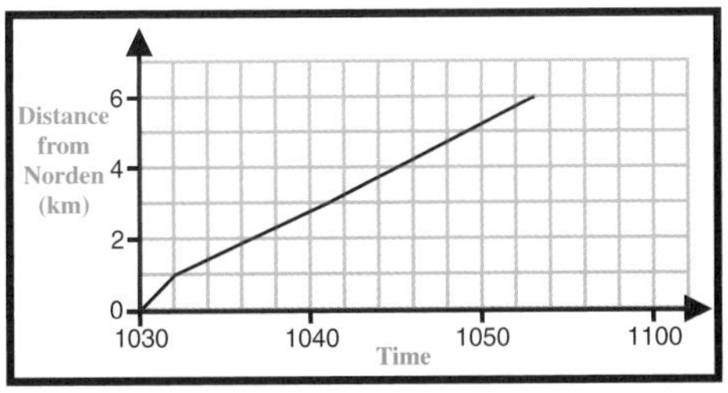

(b) Between Norden and Corfe Castle.

(c) Times and distances are joined by straight lines to show average speed.

26. 5 cm

27. £120 per month

28. (a) $\frac{17}{20}$ (b) Beth £60, Lucy £20

29. (a) $x + (2x - 1) + 3x = 41$

(b) $x = 7$ Numbers on cards: 7, 13, 21

30. (a) $x = 70°$ (alternate angles)

(b) $\angle APQ = \angle ARS = 60°$

$\angle PAQ = 180° - (60° + 70°) = 50°$

$y = 50°$, as $y = \angle PAQ$ (vert. opp. ∠'s)

31. $2^3 \times 3^2$

32. (a) Enlargement, scale factor $\frac{1}{2}$, centre (0, 1)

(b) Rotation, 90° anticlockwise, about (1, −1)

33. (a) 1 : 8 (b) 1.12 m

(c) Greatest: 21.5 cm, least: 20.5 cm

34. (a) Missing entries are: 5, −1, 1

(b)

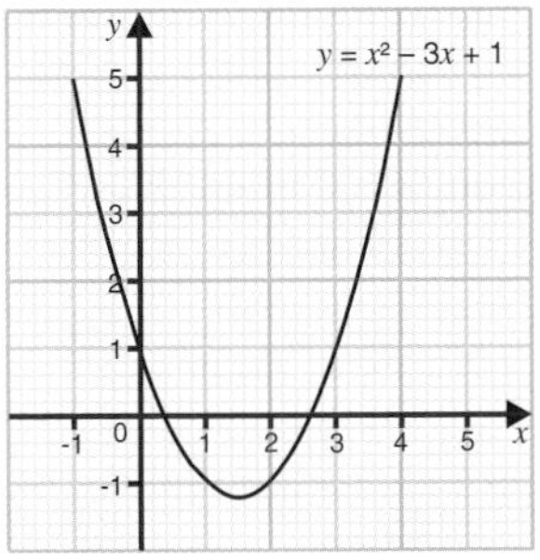

(c) $y = -1.25$

(d) $x = 0.4$ or $x = 2.6$

35. $V = 3 \times 4 \times 4 \times 8 = 384\,\text{cm}^3$

36. (a) $3(2x + 5)$ (b) $x = 8.5$

37. (a) $t = \dfrac{W - 3}{5}$ (b) m^5

38. $\sqrt{a^2 + b^2}$ and $4(a + b + c)$

Both have dimension 1.

Exam Practice

Calculator Paper Page 116

1. (a) −7, 0, 3.5, 5, 10

(b) 8 thousands, 8000

2. (a) (i) Missing entries are: 12, 16, 20

(ii) 400. Four times pattern number.

(b) **10.** Take next whole number from the last term. **16 − 6 = 10**

(c) $\mathbf{6^2 - (2 \times 6) + 1 = 25}$

3.

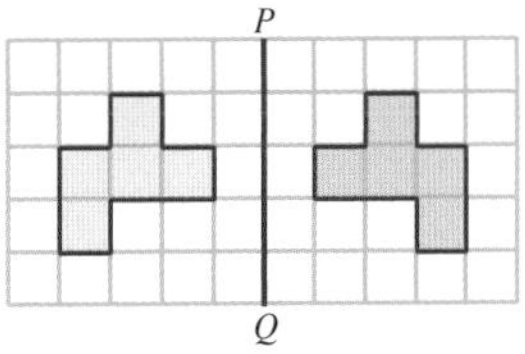

4. (a) **evens**

(b) **unlikely**

5. (a) 2 hours 35 minutes

(b) 10 minutes (c) 1220

6. Turns through 180° in a clockwise direction.

7. £86

8. **A** and **D**

9. (a) 32 (b) 17 (c) 24

10. (a) 50% (b) 25% (c) 8

11.

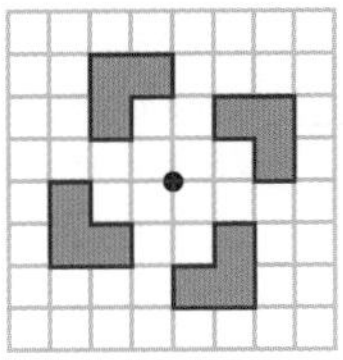

12. (a) £219 (b) 70 hats

13. (a) 4 g

(b) (i) $n = 4$

(ii) $m = 3$

(c) 18

(d) −1

14. $\frac{1}{3}$, 40%, 0.5

15. (a)

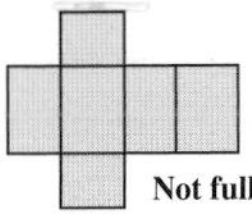

Not full size

(b) 59 cm²

16. £281.60

17. (a) −9 (b) 5

18. (a) (i) 146.41 (ii) 6.78 (iii) 25

(b) … ***six***

19. (a) 36 (b) 12.5%

20. (a) 8 gallons (b) 48 kilometres

(c) 180 miles (d) £1080

21. 24.1 kg

22. (a) 0815 (b) (i) Between 0820 and 0835 Steepest gradient

(ii) 12 km/h

23. (a) **150 g** (b) **100 g**

24. 27%

25. (a) 1260°

(b) $a = 33°$, $b = 32°$, $c = 65°$

26. (a) (i) $x^2 - 3x$ (ii) $13y + 5$

(b) (i) $3(2a + 3)$ (ii) $b(2b + 1)$

(c) (i) $x = 100$ (ii) $x = -4.5$

27. (b) 2.15 km, 295°

28. (a) 30 (b) $4 \leqslant t < 6$

(c) 4.13 hours

29. (a) 7.35 cm² (b) 72.03 cm³

(c) 6.94 g/cm³

30. $x = 3.8$

31. 61.1%

32. 312 (or 310)

33. (a) −2, −1, 0, 1 (b) $x < \frac{1}{2}$

Index

12-hour clock . . . 17
24-hour clock . . . 17
3-dimensional shapes . . . 57, 67, 68
3-figure bearings . . . 63

a

accuracy . . . 5, 80
accurate drawing . . . 55, 59, 71
acute angle . . . 53
acute-angled triangle . . . 55
adding algebraic terms . . . 31
addition . . . 1, 3, 8, 10, 31
addition of decimals . . . 3
addition of fractions . . . 10
addition of negative numbers . . . 8
algebra . . 31, 33, 34, 36, 38, 45, 46
algebraic expressions . . . 31, 36
allied angles . . . 53
alternate angles . . . 53
angle bisector . . . 71
angles . . . 53, 55, 63
angles at a point . . . 53
approximation . . . 5
arc . . . 65
area . . . 55, 59, 65, 67, 68, 80
area of a circle . . . 65
area of a triangle . . . 55
ascending order . . . 1
average speed . . . 23
averages . . . 93, 94
axes of symmetry . . . 57

b

back to back stem and leaf diagrams . . . 96
balance method . . . 33
bar charts . . . 91
bar-line graphs . . . 91
basic hourly rate of pay . . . 19
bearings . . . 63
'best buy'. . . 17
biased results . . . 88
bisector of an angle . . . 71
brackets . . . 31, 34

c

cancelling fractions . . . 10
capacity . . . 80
centre of enlargement . . . 76
centre of rotation . . . 73
changing decimals to fractions . . 3
changing decimals to percentages . . . 15
changing fractions to decimals . . 10
changing fractions to percentages . . . 15
changing percentages to decimals . . . 15
changing percentages to fractions . . . 15
changing the subject of a formula . . . 36
changing units . . . 80
checking answers . . . 5
chord . . . 65
circle . . . 65, 68
circumference . . . 65
circumference of a circle . . . 65
class intervals . . . 88
classes . . . 88
collecting data . . . 88
combinations of shapes . . . 67
combinations of transformations. 73
common difference . . . 38
common factors . . . 12
comparing data . . . 91, 93, 94
comparing distributions . 91, 93, 94
compass points . . . 63
compasses . . . 55, 59, 71
complementary angles . . . 53
compound measures . . . 23
compound shapes . . . 67
congruent shapes . . . 57
congruent triangles . . . 57
constructions . . . 55, 59, 63, 71
continuing a sequence . . . 38
continuous data . . . 88, 98
continuous measures . . . 88
conversion graph . . . 42
converting units of measurement . . . 80
coordinates . . . 40
correlation . . . 101
corresponding angles . . . 53
cube . . . 67
cube number . . . 13
cube roots . . . 13
cuboid . . . 57, 67, 68
cylinder . . . 68

d

data . . . 88
data collection sheets . . . 88
databases . . . 88
decimal places . . . 3, 5
decimal point . . . 3
decimals . . . 3, 5, 10, 15
degree of accuracy . . . 5
degrees . . . 53
denominator . . . 10
density . . . 23
descending order . . . 1
diameter . . . 65
dimensions . . . 80
direct proportion . . . 21
direction . . . 63
discrete data . . . 88
discrete measures . . . 88
distance-time graphs . . . 42
dividing algebraic expressions . . 31
dividing by 10, 100, 1000 . . . 1
dividing by multiples of 10 . . . 1
dividing decimals . . . 3
dividing decimals by powers of 10 . . . 3
division . . . 1, 3, 8, 10, 13, 31
division of fractions . . . 10
division of negative numbers . . . 8
double inequalities . . . 45
drawing angles . . . 53, 63
drawing solid shapes . . . 67
drawing triangles . . . 55

e

edges . . . 67
elevations . . . 67
enlargement . . . 76
enlarging shapes . . . 76
equally likely outcomes . . . 103
equation of a straight line . . . 40
equations . . . 33, 34, 40, 46
equidistant . . . 71
equilateral triangle . . . 55
equivalent fractions . . . 10
equivalent ratios . . . 21
estimate of mean . . . 93
estimating probability . . . 103
estimating quantities . . . 80
estimation . . . 5
events . . . 103
exchange rates . . . 17
expanding brackets . . . 31, 34
expressions . . . 31, 36
exterior angle of a triangle . . . 55
exterior angles . . . 55, 61
exterior angles of a regular polygon . . . 61

f

faces . . . 67
factorising . . . 31
factors . . . 12, 31

foreign currency 17
forming equations 34, 36
formulae 36, 80
fractional scale factors 76
fractions 3, 10, 15
frequency 88, 91
frequency distribution 88, 93
frequency distribution tables 88, 93
frequency polygons 98
frequency tables 88
function 40

g

gradient 40, 42
graphical solution of equations 40, 46
graphs 40, 42, 46
greater than or equal to, ⩾ 45
greater than, > 45
grouped data 88, 93, 98
grouped frequency distribution 88, 93, 98
groups 88, 93

h

hexagon 61
highest common factor 12
histograms 98
hourly rate of pay 19
household bills 19
hypotenuse 78
hypothesis 88

i

image 73
imperial units of measurement .. 80
improper fractions 10
income tax 19
index form 12, 13
indices 12, 13
inequalities 45
intercept 40
interest 19
interior angles of a polygon 61
isometric drawings 67
isosceles trapezium 59
isosceles triangle 55

k

kite 59

l

large numbers 13
least common multiple 12
length 80
less than or equal to, ⩽ 45
less than, < 45
linear function 40
line graphs 98
line of best fit 101
line of symmetry 57
line segments 53
linear sequence 38
listing outcomes 104
loci 71
locus 71
long division 1
long multiplication 1

m

making 3-dimensional shapes .. 67
maps 63
mass 23, 80
maximum value 46
mean 93, 94
measurement 5, 80
measuring angles 53, 63
median 93, 94
metric units of measurement ... 80
midpoint of a class 93
minimum value 46
mirror line 73
misleading graphs 98
mixed numbers 10
mode 91, 93, 94, 96
money 3, 17, 19
multiples 12
multiplication 1, 3, 8, 10, 31
multiplication of fractions 10
multiplication of negative numbers 8
multiplying algebraic expressions 31
multiplying by 10, 100, 1000, ... 1
multiplying by multiples of 10 1
multiplying decimals 3
multiplying decimals by powers of 10 3
multiplication tables 1
mutually exclusive events 103

n

negative correlation 101
negative numbers 8
nets 67
*n*th term of a sequence 38
number line 8, 45
number patterns 38
number sequences 38
numbers 1, 3, 8, 12, 38
numerator 10

o

obtuse angle 53
obtuse-angled triangle 55
octagon 61
order of operations 1
order of rotational symmetry ... 57
ordering decimals 3
ordering numbers 1, 8
outcomes 103, 104
overtime rate of pay 19

p

π 65
parallel lines 53
parallelogram 59
pay 19
pentagon 61
per cent 15
percentage change 15
percentage decrease 15
percentage increase 15
percentages 15, 17, 19
perfect correlation 101
perimeter 55, 59
perpendicular bisector of a line . 71
perpendicular from a point on a line 71
perpendicular from a point to a line 71
perpendicular height 55
perpendicular lines 53
personal finance 19
pictogram 91
pie charts 96
place value 1, 3
planes of symmetry 57
plans 67
polygon 61
population density 23
positive correlation 101
possibility space diagram 104
power form 12
powers 12, 13, 31
primary data 88
prime factors 12
prime numbers 12
prism 68
probability 103, 104
probability experiments 103
probability scale 103
product of prime factors 12
proportion 21
protractor 53, 55, 59
Pythagoras' Theorem 78

q

quadratic equations 46
quadratic function 46
quadrilateral 59, 61
qualitative data 88
quantitative data 88
questionnaires 88

r

radius 65
random events 103
range 91, 93
ratio 21
reading numbers 1
reading scales 80
rearranging formulae 36
reciprocals 13
rectangle 59, 67
recurring decimals 10
reflection 73
reflex angle 53
regular polygon 61
regular tessellations 61
relative frequency 103
removing brackets:
by expanding 31, 34
rhombus 59
right angle 53
right-angled triangle 55, 78
rotation 73
rotational symmetry 57
rounding 5
rules of indices 13, 31

s

sample 88
savings 19
scale drawing 63
scale factor 76
scalene triangle 55
scales on maps and plans 63
scatter graphs 101
secondary data 88
sector 65
segment 65
sequences of numbers 38
shapes . . 55, 57, 59, 61, 65, 67, 68
sharing in a given ratio 21
short division 1
short multiplication 1
significant figures 5
similar figures 76
Simple Interest 19
simplifying expressions 31
simplifying fractions 10
simplifying ratios 21
sketch diagrams 55
slope 40
small numbers 13
solid shapes 67, 68
solving equations 33, 34, 40
solving equations by
trial and improvement 34
solving equations by
working backwards 33
solving equations graphically . . 46
solving equations:
by inspection 33
solving equations:
using the balance method . . . 33
solving inequalities 45
solving quadratic equations 46
speed 23, 42
spread 93
square 59, 61, 67
square numbers 12, 38
square roots 12
standard index form 13
statistical bias 88
statistical diagrams 91, 96,
98, 101
stem and leaf diagrams 96
straight line graph 40
subject of a formula 36
substituting 36, 40
subtracting algebraic terms 31
subtraction 1, 3, 8, 10, 31
subtraction of decimals 3
subtraction of fractions 10
subtraction of negative numbers . 8
sum of angles in a triangle 55
sum of the angles of
a quadrilateral 59
sum of the exterior angles of
a polygon 61
sum of the interior angles of
a polygon 61
supplementary angles 53
surface area 67, 68
symmetrical 57
symmetry 57

t

table of values 40, 46
tally marks 88
tangent 65
tax 19
taxable income 19
tax allowance 19
temperature 8
terms 31
terms of a sequence 38
tessellations 61
Theorem of Pythagoras 78
three-dimensional
shapes 57, 67, 68
three-figure bearings 63
time 17
time series 98
timetables 17
'top heavy' fractions 10
transformations 73
translation 73
transversal 53
trapezium 59
trend 98, 101
trial and improvement 12, 34
trials 103
triangle 55, 57, 61, 67, 78
triangular numbers 38
two-dimensional drawing of
3-dimensional shapes 67
two-way tables 88
types of triangle 55

u

unitary method of sharing 21
units of measurement 80
using a calculator 5, 12, 13
using formulae 36

v

VAT 17
variables (statistical) 88
vector 73
vertex 53
vertically opposite angles 53
vertices 67
volume 23, 67, 68, 80

w

wages 19
whole numbers 1, 5
writing equations 34
writing numbers 1

x

x coordinate 40

y

$y = mx + c$ 40
y coordinate 40

z

zero correlation 101